A PRACTICAL GUIDE TO INSPECTING

ELECTRICAL

By Roy Newcomer

D1310676

PGIEL 01/07

CONTENTS

INTRODUCTION

My background includes many years in construction and several more as the owner of a Century 21 real estate franchise. In 1989, I started a home inspection company that has grown larger than I ever imagined it could. Training my own staff of inspectors to the highest inspection standards has led to my teaching home inspection seminars across the country and developing study courses, books, and videos for home inspectors. The American Home Inspectors Training Institute was founded as a result of my desire to share this experience and knowledge in home inspection.

The *Practical Guide to Inspecting* series is intended for both beginning and experienced home inspectors. So if you're studying home inspection for the first time or are using the materials as a refresher, these guides should be of assistance to you.

I've created these guides to include all aspects of home inspection. Not only a broad technical background in home systems, but the other things you need to know in order to perform a *good* inspection of those systems. They lay out technical information, guidelines for the inspection, how-to instructions for inspecting system components, and the defects, deficiencies, and problems you'll be looking for during the inspection. I've also included some advice on how to report your findings to the home inspection customer.

I've been a member of several professional organizations for a number of years, including ASHI® (American Society of Home Inspectors), NAHI™ (National Association of Home Inspectors), and CREIA® (California Real Estate Inspection Association). I am a great supporter of those organizations' quest to promote excellence in home inspection.

I do encourage you to follow the standards of the organization to which you might belong, or any state regulation that might take precedent over the standards used here. Use the standards in this book as a general guide for study and apply the standard or state regulation that applies to you.

The inspection guidelines presented in the Practical Guides are an attempt to meet or exceed standards and regulations as they exist at the revision date of the guides.

There's a lot to learn about home inspection. For beginning inspectors, there are some *hands-on exercises* in this guide that should be done. I'm a great believer in learning by doing, and I hope you'll try them. There are also some of my *personal inspection stories* to let you know what it's really like out there.

The *inspection photos*, located in this text can also be found at www.ahit.com/photos, are referenced in the text. You'll read the story about each one as you go along. Be sure to watch for my *Don't Ever Miss* lists. I've included them to alert home inspectors to report those defects (if found during the inspection) in the inspection report. If missed, these items are often the cause for lawsuits later. Finally, to help you see how you're doing as you study this guide, I've included some *worksheets*. The answers are given for each one for self checking. Give them a try. Checking yourself can help you lock important information in your mind. There's also a *final exam* that you can complete and send in to us. Many organizations and states have approved this book for continuing education credits. Submit the exam with the required fee if you need these credits.

In total, the *Practical Guide to Inspecting* series cover all aspects of the general home inspection. Each guide covers a major aspect of the inspection, as their titles show:

Electrical
Exteriors
Heating and Cooling
Interiors, Insulation, Ventilation
Plumbing
Roofs
Structure

If you are interested in other titles in the series, please call us at the American Home Inspectors Training Institute to order them. Call toll free at 1-800-441-9411.

Roy Newcomer

INSPECTING
ELECTRICAL

Chapter One

THE ELECTRICAL INSPECTION

The electrical inspection of the home includes the inspection of the following components:

- The service entrance from the masthead and meter box to the main panel
- The main panels and subpanels
- Branch circuit wiring
- Junction boxes, outlets, and fixtures

All visible components of the electrical system, as listed above, will be evaluated during the inspection. We stress the word *visible*. The home inspector removes the cover to the main panel and subpanels to view the inside but is not expected to determine what is occurring behind drywall or inside the ceilings. All standards of practice state that the electrical inspection is a **visual inspection, not a code inspection.**

Electrical installations for homes and commercial buildings are governed by the **National Electrical Code** (NEC). The NEC, which has been around for decades with revisions every 3 years, is considered to be the authority on safe wiring practices. Requirements differ for residential and commercial properties and, generally, are less strict for private homes than for buildings meant for public occupancy. States and local communities are not required to accept the NEC as their standard, although most follow it to some extent. Some locations may adopt each new NEC update, while others may follow an earlier revision.

The NEC doesn't require electrical systems in older homes to be updated with each new revision of the code. For example, old knob-and-tube wiring cannot be installed any longer, but the code does not require existing knob-and-tube wiring to be replaced. The home inspector may see remodeled buildings wired under both old and revised sections of the code.

Although the home inspector performs only a visual examination of the electrical system and not a code inspection, he or she should be familiar the National Electrical Code.

The most important aspect of the electrical inspection is to be on the alert for **safety hazards**. Many house fires are the result of faulty electrical systems. The home inspector can play an important role in protecting new homeowners from such dangers.

Guide Note

Pages 1 to 8 outline the content and scope of the electrical inspection, according to most standards of practice. It's an overview of the inspection, including what to observe, what to describe, and what specific actions to take during the inspection. Study these guidelines carefully.

These pages also present some special cautions about inspecting the electrical system. Please read them and be aware of the potential of harming yourself or interrupting power flow to the house.

For Your Library

You might want to have a copy of the National Electrical Code for your own information and study. You can obtain a copy from the International Association of Electrical Inspectors or at most larger bookstores.

ELECTRICAL INSPECTION

- Service entrance from masthead to main panel

- Main panels and subpanels

- Branch circuit wiring

- Junction boxes, outlets, and fixtures

Inspection Guidelines and Overview

Study these guidelines carefully.

Electrical System	
OBJECTIVE	To identify major deficiencies in the electrical system.
OBSERVATION	Required to inspect and report: • Service entrance conductors • Service equipment, grounding equipment, main overcurrent device, main and distribution panels • Amperage and voltage rating of the service • Branch circuit conductors, their overcurrent devices, and the compatibility of their ampacities and voltages • Polarity and grounding of all receptacles within 6 feet of interior plumbing fixtures, all receptacles in the garage or carport, and on the exterior of inspected structures • Service type as being overhead or underground • Location of main and distribution panels Not required to observe: • Low voltage systems • Smoke detectors • Telephone, security, cable TV, intercoms, or other ancillary wiring that is not part of the primary electrical distribution system
ACTION	Required to: • Operate a representative number of installed lighting fixtures, switches and receptacles located in the house, garage, and exterior. • Operate GFCI's. • Report any observed aluminum branch circuit wiring. Not required to: • Insert any tool, probe, or testing device inside the panels. • Test or operate any overcurrent device excepts GFCI's. • Dismantle any electrical device or control other than to remove the covers of the main and auxiliary distribution panels.

The table on page 2 provides a good outline of the guidelines that govern the inspection of the electrical system. Not every single detail is presented in the table, but enough of an overview is presented to have a good idea of what is inspected and what is not required to be inspected. Here is an overview of the electrical inspection as stated in the most standards of practice.

The table on page 2

- **Service entrance:** The home inspector determines if the electrical service coming to the house is **overhead or underground**. If over head, the home inspector observes overhead wire for the proper height above the driveway, above the ground near the house, and, if attached to the masthead on the roof, proper height above the roof. The masthead, outside conduit or cable running to the meter, and the meter itself are inspected for **secure attachment** to the house. The inspector may find, for example, that these items are not securely attached, rusting, bent or broken, or not properly waterproofed.

  The home inspector identifies the service conductor **material** used as either copper, aluminum, or copper-clad aluminum and determines the **amperage** of the service supplied to the home (for example, 60-amp, 100-amp, or 200-amp service) and its **voltage rating** (for example, 240 volts). The home inspector locates the **main overcurrent device** (the main disconnect which may be found next to the meter or inside the main panel) and determines the compatibility of wiring on both sides.

  The inspector also determines if the electrical system is **grounded** by noting the presence of the grounding conductor in the main panel and its termination to a water pipe, foundation footing (Ufer) or driven rod ground(s).

- **Main panel and subpanels:** The home inspector is required to locate the main and subpanels and, if conditions are safe, to **remove the panel covers** in order to inspect their interiors. However, note that the inspector is **not required to dismantle any other electrical device or control in the electrical system** other than removing these covers. Also note that the inspector is **not required to insert any tool, probe, or testing device inside the panels**.

Guide Note

There are many electrical terms used on this page and the next. Don't worry about them at this time. They'll all be defined when the topics are discussed in detail later in this book.

The home inspector again checks for compatibility of components — of cables to main disconnect, of fuses or breakers with incoming and outgoing wire size. The home inspector trips the test button on GFCI breakers (ground fault circuit interrupter) in the main panel. However, the inspector is **not required to test or operate any overcurrent device** (main disconnect, fuse, or breaker) **except GFCI's.**

Some of the defects noted in the main panel can include over sized fuses or breakers, double tapping, melted insulation, corrosion, arcing, rusting inside the box indicating the presence of water or condensation, and damage to the panel box itself such as a missing cover or loose attachment to the wall.

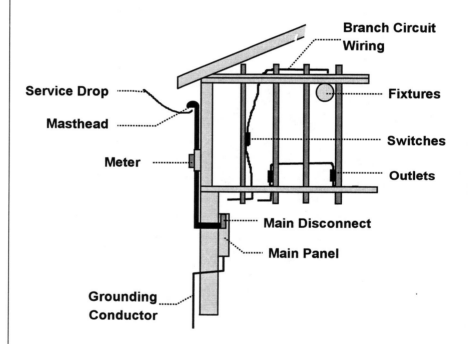

NOTE: The home inspector is **not required to inspect low voltage systems**. Low voltage systems, popular in the 1950's, are defined as one that features boxes of relays, unusual switches, and one or more switching panels from which all the lights in the house could be controlled. They allowed unusual lighting effects and operating flexibility. These relay boxes are usually tucked away in remote corners and can be hard to reach. The home inspector should, however, test to see if wall switches work.

- **Branch circuit wiring:** This is the wiring that carries electricity from the main panel to the fixtures and appliances in the house. The home inspector identifies the **type** of branch circuit wiring as copper or aluminum and cabling as armored cable, Romex®, conduit, or knob-and-tube wiring. The **condition** of the wiring and its proper installation are inspected where possible throughout the living area of the house, basement, attic, garage, and any exterior wiring.

  Findings may be poorly secured wiring, damaged insulation on wiring, installation too close to heating ducts and hot water piping, handyman or extension-cord wiring, and wire which is undersized for the appliance it serves. The home inspector is **required to report any observed solid conductor aluminum branch circuit wiring**. This is because of fire safety problems occurring with aluminum wiring, which was installed during the mid-1960's to mid-70's. We also suggest the home inspector recommend an electrician be called in to inspect any knob-and-tube wiring found, since it was installed at least 75 years ago — 1920's and 1930's.

  NOTE: The home inspector is **not required to inspect telephone, security, cable TV, intercoms, or other ancillary wiring** in the house that is not part of the primary electrical distribution system.

- **Junction boxes, outlets, and fixtures:** The home inspector inspects all outlets within 6' of water (as in the kitchen and the bathroom) for **polarity and grounding**. Using the GFCI tester, the inspector tests and trips **all GFCI's** in the house, garage, kitchen, and on the exterior of the house. All wall switches should be operated in the house and at least one outlet in each room, hall, and stairway areas. The home inspector inspects **junction boxes** in exposed areas such as garage, attic, and basement, looking for boxes without covers, improperly made splices, and other unsafe wiring practices.

Take a tour around your own home to identify the main components of the electrical system. Look at the service entrance components from outside the house. Locate the main panel. From the main panel, trace the branch wiring, noting junction boxes, outlets, ceiling fixtures, where wiring goes up into the house, and so on. Try to visualize a map of wiring for the house. Please don't touch the main panel or any wiring at this time.

Personal Note

"One of my inspectors had removed the main panel cover and was inspecting the wiring. His customer, standing close by, asked, 'What's this?' and suddenly reached up and pointed into the panel, touching a breaker. The customer received a shock that threw him back a few feet.

"Although the inspector was not the cause of the problem, I consider it his fault that the customer received the shock. The inspector should have had the customer stand far enough away so he couldn't reach the panel."

Roy Newcomer

Inspection Equipment

The 2 most important tools you'll need for the electrical inspection are a **GFCI tester** and the **neon bulb tester**. The GFCI tester is a little 3-pronged device with 3 lights on its face. It's used to test GFCI outlets. The neon bulb tester is designed specifically for testing 2-slot outlets. But both testers can be used to test 3-slot outlets for power, grounding, and polarity. We'll explain how to use these tools later in this book.

Inspection Tools

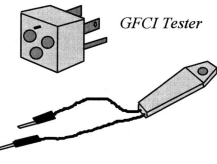

GFCI Tester

Neon Bulb Tester

For the electrical inspection, the home inspector should also have a high-power **flashlight** for inspecting wiring in dark areas and **screwdrivers** for removing the panel cover. **Binoculars** are useful when inspecting the overhead service and the masthead from the ground. There are other devices available that some inspectors purchase, such as the voltmeter and the ammeter, which can be helpful but are not necessary for the general home inspection.

The beginning home inspector may want to purchase short lengths of various size and types of **electrical wires and cables** to have on hand for identification purposes.

Inspection Concerns

We all know that electricity can be dangerous. We know that water and electricity make a deadly combination. And we know that wiring generating heat and sparks is something not to grab hold of. Don't we? Well, home inspectors can get careless, and getting shocked and thrown across the basement is not unheard of. Always keep in mind the dangers that may be present in an electrical system. Have the proper respect for the power that's present and for its ability to harm you. Be smart.

These are some general rules about the electrical inspection that the home inspector should follow.

- **Keep your customer safe.** Of greatest concern during the electrical inspection is the customer's safety. The home inspector is responsible for the well-being of the customer. Good home inspection practice is to have the customer present during the home inspection, and that does mean during the inspection of the electrical system too.

  However, we suggest the home inspector have the customer stand back a **safe distance** during this part of the inspection — especially at the main panel, where more things can go wrong. Put yourself between the main panel and the customer. People are just too unpredictable. They can intentionally touch things they shouldn't or bump into things accidentally. As you perform each step in the electrical inspection, always be aware of where the customer is and take measures to protect his or her safety. Tell your customer why you're requesting that he or she stand back. Explain that it's for safety reasons.

- **Protect yourself from harm.** The home inspector should always exercise caution when performing all the steps in the electrical inspection.

  Working at the **main panel** can be especially dangerous. There are times when the main panel cover should not be removed and the panel should be left uninspected — if there is water on the floor under the panel box, if the panel box is warm to the touch, if the home inspector can hear arcing inside the panel, and if there is water leaking into the panel. In these cases, simply stay away and inform the customer that a licensed electrician will need to inspect the panel.

- **Don't do any electrical work.** The home inspector is present to report deficiencies in the electrical system, not to perform any electrical work. The home inspector is not there to change fuses or fix reverse polarity in outlets. The safest thing to do it stick to the standards of practice and to do nothing more.

- **Don't turn off the power.** The electrical inspection does not include turning off the main disconnect or operating the breakers. An interruption to the power flow to the

---

IMPORTANT RULES

- Keep your customer safe.
- Protect yourself from harm.
- Don't perform any electrical work.
- Don't turn off the power.
- If you have to check pull-out fuses, check with the owner before interrupting power.

Personal Note

"This is a case of not thinking and not putting 2 and 2 together. Electricity plus water equals danger.

"One of my inspectors-in-training actually put his hand into a sump pump to test its operation. He got shocked and thrown over onto his back. Fortunately, it only happened that one time. What a way to learn a lesson."

Roy Newcomer

house can cause all sorts of problems. Circuit breakers may not reset or may malfunction. Appliances or equipment in the house can be affected by momentary loss of power. Clocks have to be reset, burglar alarms may be triggered, and computer data can be lost.

- **Check with the owner before interrupting power:** If a home has a 60-amp panel with fused pull-out blocks, the home inspector would have to pull out the block to read fuse size. This, of course, would interrupt power. The home inspector should check with the owner first to see if interrupting the power will cause any problems. Testing GFCI's will also interrupt power, but it's unusual to have the kitchen or bathrooms on the same circuit with other rooms where sensitive equipment may be plugged in.

Chapter Two
BASIC ELECTRICITY

As a home inspector, it's important to understand the basic principles of electricity. An inspector's liability is very high and very expensive where the electrical system is concerned. House fires can start from a fault in the system. The home inspector wants to be able to point out safety hazards during the inspection.

Volts, Ohms, Amps, and Watts

Electricity is supplied to homes from a power plant in an **alternating current**. It doesn't flow in one direction like water through a pipe to a house. An alternating current means that electrons are made to move back and forth at a frequency of 60 cycles per second. An alternator at the power plant creates the movement of electrons, and it's really this back-and-forth *movement* that travels along power lines to the lights and appliances in the home.

- **Resistance:** The alternating current of electricity is channeled through **conductors** such as copper wires. As the current alternates back and forth through conductors it experiences **resistance** (opposition or something like friction) that makes heat. Resistance is expressed in **ohms**.

  Good electrical conductors have a **low resistance**. Silver is a good conductor, as is copper. So is aluminum, although aluminum has a higher resistance than copper. Therefore, a larger size aluminum wire would be needed to carry the same current as a smaller size copper wire. Conductors are used to channel electricity where we want it to go. By the way, water is an excellent conductor of electricity, which is why water and electricity are so dangerous together.

  Materials with a **high resistance** are poor conductors of electricity. They're called **insulators** and include materials like wood, rubber, ceramic, and most plastics. Insulators are used to keep electricity from going where we *don't* want it to go. Air is an excellent insulator.

  A light or appliance such as a stove or refrigerator, in electrical terms, is said to be a **load**. Each load has its own resistance.

Guide Note

Pages 9 to 14 present an overview of electrical concepts important to the understanding of every home inspector. The application of these concepts to the practical matter of inspecting electrical systems will be studied later, beginning on page 16.

Definitions

Current is the flow of electricity and is expressed in amps. In an alternating current, electrons move back and forth at a frequency of 60 cycles per second. Ampacity refers to the amount of current that can safely pass through a conductor.

Resistance is the opposition offered by a material when a current passes through it. Resistance is expressed in ohms.

A good conductor is a material that offers a low resistance to an electric current flowing through it. An insulator is a material that offers a high resistance.

The electromotive force, or voltage, is what drives the current of electrons through a given resistance. It is expressed in volts.

Power is the heat produced by the flow of current through a given resistance. Power is expressed in watts, kilowatts, or British Thermal Units (BTU's).

- **Electromotive force:** The electromotive force, or the **voltage**, is the potential energy of the electrical system. The electromotive force is expressed in **volts**. This force is what drives a current of electrons through a given resistance. Homes today are supplied with a 120/240 volt electrical system which can provide 240 or 120 volts. That is, within the house, the voltage is broken down for most appliances to 120 volts and available for others such as central air conditioners as 240 volts. Houses received 110/220 volts in the 1950's and 115/230 volts in the 1970's. The change over the years is an improvement of the electrical service available to homeowners.

- **Current:** The current is the flow of electricity that results when the electromotive force is applied across a given resistance. Current is expressed in **amps** or amperes. This unit of measure refers to the number of electrons flowing past any given point in the circuit during a given time.

  The flow of electricity generates heat. The more amps flowing through a wire, the hotter a wire will get. If a wire gets too hot, it is no longer safe. Therefore, different sized wires are required for different amp services. The term **ampacity** refers to the number of amps that may be *safely* pushed through a conductor. An ampacity rating is given to wire according to the temperature rating of its insulation, the size of the wire, and the metal it's made of.

  Electrical current is what electrocutes people, not the voltage or wattage. A current of less than one amp is capable of killing someone.

  Older homes were provided with a 30-amp or 60-amp service. Most homes today require at least a 100-amp service and more typically 200 amps if the home is of newer construction

- **Power:** Power can be defined as the output of the electrical system or its ability to do work. Power is the amount of heat produced when the current moves through the resistance of a light or appliance — when electrical energy is changed into heat energy. Power is expressed in **watts**, and 1000 watts is called a **kilowatt**. One watt is equal to 3.4 **BTU's** (British Thermal Units).

The units of measure of electromotive force, resistance, current, and power are basically defined in terms of each other.

	Measure	Letter	Definition
Electromotive force	Volts	E	Force that drives current through a given resistance
Resistance	Ohms	R	Opposition of a material to the flow of electricity through it
Current	Amps	I	Flow of electricity
Power	Watts Kilowatts BTU's	W kW BTU	Heat produced by flow of electricity through a given resistance

ELECTRICAL FORMULAS

Volts = Amps x Ohms

Watts = Volts x Amps

An electromotive force of 1 volt will push a current of 1 amp through a resistance of 1 ohm. This relationship is expressed in the following formula:

$$E = I \times R$$

Electromotive force = Current × Resistance

Or you can say that *volts equal amps times ohms*. Using the formula above, you could calculate the force it would take to push 2 amps of current through 60 ohms of resistance. The force would equal 2 × 60, or 120 volts. You can manipulate the formula to find the resistance (R = E/I) or to calculate the current (I = E/R).

The power of 1 watt is produced by a current of 1 amp pushed by an electromotive force of 1 volt. This relationship is expressed in another formula:

$$W = E \times I$$

Power = Electromotive Force × Current

This can be thought of as *watts equals volts times amps*. You can use the formula to calculate the power available with a 240-volt service and 100-amp main fuses. The power would equal 240 × 100, or 24,000 watts. You could find force (E = W/I) or current (I = W/E) by changing the formula around.

Definitions

An <u>electrical circuit</u> is a complete path of an electric current. An <u>overload protection device</u> is a fuse or breaker which will break the circuit when it overloads.

The <u>neutral wire</u>, for purposes of our discussions, is the grounded center line coming into the house from the power company. The 2 <u>hot wires</u> coming from the power company are each charged with 120 volts.

Circuits

A circuit is a complete path of an electric current. Electricity doesn't flow like water through a pipe that ends at a faucet, and it doesn't get used up like the water coming out of the faucet. Electrons flow, in that back-and-forth movement, through a light or appliance and back to its source. The circuit is actually a **circular path**.

The alternator at the power company generates the flow of electricity over high voltage power lines. A **transformer** on a utility pole reduces the voltage to its appropriate level for home use and provides a third line. This center line is grounded and is called **neutral** (equal and opposite currents cancel each other out). The outer 2 lines remain **hot** or **live**, each with a charge of 120 volts.

These 3 lines provide the home with both 120-volt and 240-volt circuits. Loads such as lights and small appliances connected **between a hot wire and the neutral complete a 120-volt circuit**. Loads such as central air conditioners or a kitchen stove connected **between the two hot wires complete a 240-volt circuit**.

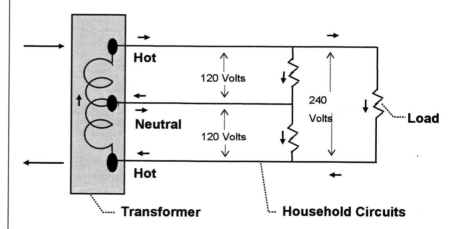

In the main electrical panel within the house, these 3 circuits can be connected to as many **branch circuits** as desired for household use, provided that each branch is protected by an **overload protection device**. Electrical wire that can safely carry the current is used. Overload protection devices — **fuses and breakers** — shut off power by breaking the circuit when more current is flowing through the circuit than the wire can handle.

Each 120-volt household circuit, for example, may have a 15-amp fuse or breaker. This means the circuit is capable of supplying 1800 watts of power (W = E × I, or 120 × 15). If a 1200-watt toaster is connected to the circuit, a current of only 10 amps would flow (I = W/E, or 1200 ÷ 120). However, if another 1200-watt appliance were added, a current of 20 amps would be drawn through the circuit (I = W/E, 2400 ÷ 120). The 15-amp fuse would blow or the breaker would shut off the circuit. If the protection device were not there, the wire would overheat and possibly cause a fire.

The amount of current a wire can carry depends on its diameter. The larger its diameter, the more current a wire can safely carry. Since aluminum is not as good a conductor as copper, an aluminum wire must be larger to safely carry the same current as a copper wire. Household circuits, designed to carry 15 amps of current, can be safely wired with #14 copper wire as the following table shows.

POWER AND CURRENT FOR APPLIANCES

	Watts	Amps
Stove	9600	40.0
Microwave	1500	12.5
Iron	1000	8.5
Color TV	360	3.0
Stereo	120	1.0

Following table is for branch circuits and feeder wire size:

Wire Gauge	Maximum Amperage Allowed Per NEC table 310.16	
	Copper Wire	Aluminum or Copper-Clad Aluminum Wire
#14	15 amps	-
#12	20 amps	15 amps
#10	30 amps	25 amps
#8	40 amps	30 amps
#6	60 amps	35-40 amps
#4	70 amps	45-50 amps
#3	80 amps	60 amps
#2	90 amps	70 amps
#1	110 amps	80 amps

The tables on pages 13 and 14 lay out the sizes of commonly used electrical wiring. Go to your local hardware store and buy short lengths of several gauges. Tape them on a small cardboard card then label each one. The card will come in handy when you begin inspecting electrical systems.

Service Conductors Wire Sizes		
Fuse or Breaker	Copper Wire	Aluminum Wire
100	4	2
110	3	1
125	2	1/0
150	1	2/0
175	1/0	3/0
200	2/0	4/0

Service conductor sizes are based on the wire types in NEC table 310-15(b)(6).

NOTE: **Copper-clad aluminum wire** is an aluminum wire with an outer cladding or covering of copper. The wire, viewed from the side, appears to be copper. The home inspector must look at the tips of the wire to see its aluminum center. Incoming service conductors may be copper-clad aluminum.

WORKSHEET

Test yourself on the following questions.
Answers appear on page 16.

1. According to most standards of practice, the home inspector is required to:

   A. Operate all overcurrent devices.
   B. Describe service entrance conductors.
   C. Dismantle electrical devices and controls.
   D. Observe security and intercom wiring.

2. While inspecting the service entrance, the home inspector is <u>not</u> required to:

   A. Determine if the service is overhead or underground.
   C. Inspect the masthead and meter for secure attachment to the house.
   D. Determine amperage and voltage rating.
   E. Inspect connections at the utility pole.

3. While inspecting the main electrical panel, the home inspector is required to:

   A. Test GFCI's.
   B. Turn off the main disconnect.
   C. Test and operate all fuses or breakers.
   D. Probe and test wire connections.

4. Which of the following does <u>not</u> have to be inspected during the electrical inspection?

   A. Exterior fixtures and receptacles
   B. Branch circuit wiring
   C. Polarity and grounding of receptacles within 6' of water
   D. Low voltage systems

5. What is electrical current?

   A. Flow of electricity
   B. Force that drives electrons through a given resistance
   C. Opposition of a material to the flow of electricity.
   D. Heat produced by the flow of electricity through a given resistance

6. Which of the following causes electrocutions?

   A. Electromotive force
   B. Resistance
   C. Current
   D. Power

7. Which formula expresses the correct relationship between electromotive force, current, and resistance?

   A. $R = I \times E$
   B. $I = R \times E$
   C. $E = I \times R$
   D. $E = R/I$

8. When electrons are made to move back and forth at 60 cycles per second, that's called:

   A. A 120-volt circuit
   B. An alternating current
   C. A complete circuit
   D. A transformer

9. Which of the following materials has the lowest resistance to the flow of electricity?

   A. Aluminum
   B. Ceramic
   C. Copper
   D. Rubber

10. In a typical household electrical service, which circuit shown below provides a 240-volt circuit — A, B, or C?

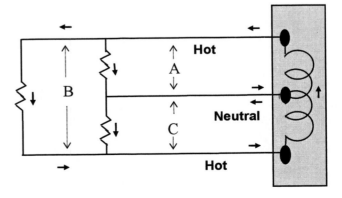

INSPECTING
SERVICE ENTRANCE

- Overhead or underground

- Clearances

- Amperage and voltage rating of service

- Type of conductors

- Condition of components

- Compatibility of components

Guide Note

Pages 16 to 34 present information about inspecting the service entrance.

Worksheet Answers (page 15)

1. *B*
2. *E*
3. *A*
4. *D*
5. *A*
6. *C*
7. *C*
8. *B*
9. *C*
10. *B*

Chapter Three

INSPECTING THE SERVICE ENTRANCE

Inspection of the service entrance portion of the electrical system begins outside where the home inspector determines whether the service is **overhead or underground**. For overhead services, the inspector checks that **clearances** above drives and walkways are the proper heights. During the inspection, the home inspector will observe the various components of the service entrance and determine the **amperage and voltage** rating of the service. The **type** of service conductors is identified. All components are inspected for their **condition** and their **compatibility** with each other.

The service entrance inspection includes inspecting the condition of the following components:

- The **service drop** from the utility pole to the masthead

- The **conduit or cable** from the masthead to the meter and from the meter to the main disconnect

- The **service entrance conductors** at the main disconnect

- The **main disconnect** whether housed separately in a service box or as part of the main panel

- The **grounding conductor** and system in general

The Service Drop

When overhead wires from a utility pole bring power to a house, it's called a **service drop**. If power is supplied underground, it's called a **service lateral**. Normally, a transformer is on the pole or outside in a pit. The service supplied to residential homes is usually 200 amps or less. For larger services, up to 600 amps, a transformer may be present inside the building. But that building would most likely be a commercial building.

The home inspector can sight along the overhead service drop with binoculars from the ground. Notice any obvious frayed wires or tree branches that may be scraping on the wires. When pointing any minor or major problems to customers, remind them that the utility company is responsible for them. One should always call the utility company for any repairs needed.

Overhead wires must meet certain clearances according to the National Electrical Code (NEC). The reason for these clearance requirements is to prevent anyone from accidentally touching the wires and to prevent vehicles from touching them. The drawing shown below indicates the minimum requirements — at least 10' above the ground and walkways, 12' above the driveway, 3' above the roof (an exception), and at least 3' from windows, doors, balconies, and decks. Wires should clear any roof ridge by 3', and with a flat roof that can be walked on, the clearance should be a minimum of 8' above the roof surface.

CLEARANCES

- 10' above yard and walkways
- 12' above driveways
- 8' above the roof with exceptions:
  - 3' above roof if slope is 4/12 or greater
  - 18" above roof if conductors don't pass over more than 4' of overhang
- 3' from windows, doors, balconies, and decks

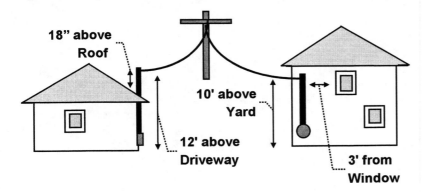

18" above Roof

10' above Yard

12' above Driveway

3' from Window

Any deviation from these NEC clearance requirements should be reported as a **safety hazard**. Instruct your customers to call the utility company to fix any problems. The responsibility may be the homeowner's or the utility company's.

The number of conductors (wires) going into the masthead is usually an indication of the **voltage rating** of the service. If only 2 wires are connected, the service is usually 120 volts. There may be 3 conductors on the service drop with only 2 wires entering and the third tied back. That indicates a 120-volt service (see drawing below). If there are 3 wires entering, it indicates 120/240-volt service.

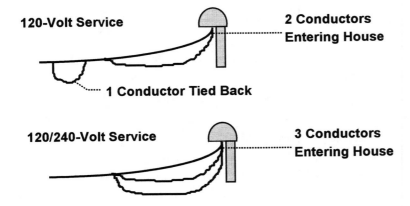

120-Volt Service

2 Conductors Entering House

1 Conductor Tied Back

120/240-Volt Service

3 Conductors Entering House

***Photo #1** shows a **240-volt overhead service** to a house (all 3 wires entering). To the left of the masthead, there is a **bracket** that supports the weight of the wires. This is needed so the weight is not supported by the clamps or wires themselves. Notice also that the incoming wires sag or go slack as they enter the masthead. This is called a **drip loop**, which causes water to run down the wires and not into the conduit. The service drop shown in this photo is okay as far as requirements go; the masthead is the proper distance from the window and from the ground (although the photo doesn't show that).*

NOTE: On rare occasions, you may see 4 conductors entering a home. This will be a **3-phase** 120/240-volt service. Single-phase power is generated at 60 cycles a second. A 3-phase service is also generated at 60 cycles but with 3 power peaks in each cycle and is usually used in industrial operations.

#1 240-volt overhead service

We've seen mastheads in unsafe locations. One example was where a deck had been added to the house without moving the masthead. The masthead was attached to the house at about shoulder height as you stood on the new deck. Any adult standing on that deck could easily reach up and touch those wires. This was an extremely dangerous situation. We reported it as a **safety hazard**.

Conduit or Cable

The overhead conductors coming from the utility pole will enter the **masthead** at the side of the house or on the roof. If the masthead is on the roof, the bracket must be raised on a **mast** standing 18" above the roof's surface (see box on page 17).

The mast must be supported so that it will not bend or break. It may be attached to the roof by guy wires or other supports. The home inspector should inspect the mast and any supports present for secure attachment and for any rust or corrosion present. Inspect the masthead itself for secure fit over the conduit or cable carrying the incoming conductors. The masthead should be firmly attached and should waterproof the service entrance conduit or cable.

INSPECTING SERVICE DROP

- Frayed or damaged overhead wires
- Improper clearances
- Determine voltage rating.

IMPORTANT POINT

The amperage of the electrical service <u>cannot</u> be determined by the size or type of the conduit bringing the conductors into the house.

The conductors coming down the outside of the house to the meter and from the meter into the house may be encased in one of the following:

- **Conduit:** Some local codes may require conduit.
- **Plastic-sheathed cable:** The NEC allows this type of cable.
- **Flexible, armored cable:** Since 1987, the NEC has allowed up to 6' of armored cable to be used outside the house. Local codes may vary.

Conduit or cable should be securely fastened to the siding with the appropriate strapping. The first strap should be present within a foot of the masthead. Other straps should be at intervals of up to 4 1/2' down the side of the house. The conduit or cable should not be loose enough to swing back and forth. There should be no vegetation touching the conduit or cable.

Sight along a flexible, armored cable to be sure it isn't worn or frayed at any point. Wear or fraying can allow water to enter the cable, causing rusting and corrosion in the main panel. Outside, an exposed neutral conductor can be a safety hazard if the grounding system is defective. Inside, water in the main panel can also be a safety hazard.

The home inspector should inspect the siding where the conduit or cable penetrates it for entrance into the house. This hole should be sealed to prevent water and air penetration.

The Meter

The **electrical meter** may be outside or inside the house. It registers electrical use in watts, either numerically or on a series of dials.

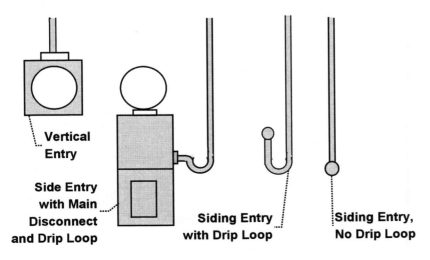

Vertical Entry

Side Entry with Main Disconnect and Drip Loop

Siding Entry with Drip Loop

Siding Entry, No Drip Loop

Personal Note

"I once inspected a house and noticed where the owner had cut the wires coming from the utility pole right at the edge of the roof. He wired them together, ran them down to the ground and into the basement window. He was stealing electricity.

"If you ever notice any tapping of wires between the pole and the meter, that means someone is cheating the utility company. That should be pointed out to your customer."

Roy Newcomer

The meter should be connected to its pan or base by a ring that is sealed by the utility company. Inform the customer if the seal is broken.

When you inspect the electrical meter, be sure to check if there is any tapping into the incoming service wires **before the meter**. Tapping lines off the incoming service drop at the roof or side of the house and tapping into the conduit of cable before the meter is illegal. It means the homeowner is stealing power from the utility company. This should be brought to the attention of your customer.

The shape of the meter base may help the home inspector in making a determination of amperage rating of the electrical service. It's not the only item to check, but it does give a clue.

- **Rectangular-base meters:** This is the type installed within the last 20 years. They're usually compatible with **200-amp services**, although there are exceptions. In 1976, an Underwriter's Laboratory ruling required all new meter bases be "continuous rated for 200 amps." Otherwise, only 80% of the actual rating would be considered. That is, a 200-amp meter base that was not continuous rated could only be used on a 160-amp service or smaller. In general, it can be assumed that a rectangular-base meter is compatible with the system it's serving, unless a major change or new panel has been installed.

- **Round-base meters:** These meters were installed 40 or more years ago. They were originally rated for **60 amps** when installed over 50 years ago. Later generations of round-base meters were rated at **100 amps**.

- **Square-base meters:** These meters were also installed 40 or more years ago. They were normally rated at **100 amps**, with later generations sometimes rated at **125 amps**.

The meter itself is a consideration in determining the amperage rating of the service. Most modern meters for single-family homes have the designation **CL200** or **200CL** somewhere on their face, indicating they are rated for up to a 200-amp service. The home inspector may find a **CL10** meter, which is a transformer-rated meter for large houses with larger electrical systems or 2 separate main panels.

Some older meters have other designations such as 15 amps on their faces. This was their test rating. These meters are only usable on systems up to 100 amps. One may occasionally find an upgraded 200-amp service that still has an old 15-amp meter plugged into a new meter base.

REMINDER: You cannot determine amperage rating on the basis of the shape of the meter base alone.

Service Entrance Conductors

The service conductors are the wires coming into the home from the utility company's meter. The home inspector is **required to identify** these conductors as copper, aluminum, or copper-clad aluminum. The materials will vary from community to community.

The size of the incoming service conductors is an indication of the **amperage rating** of the service. The home inspector can compare his or her own sample wire sizes to the service conductors at the main disconnect or the main panel. Aluminum and copper-clad aluminum wires require the same gauge for various amperage ratings (see chart below).

Amperage Rating	Copper Wire Size	Aluminum Wire Size
30 amps	#10	#8
60 amps	#6	#4
100 amps	#4	#2
125 amps	#2	#1/0
150 amps	#1/0	#2/0
200 amps	#2/0	#4/0

SERVICE CONDUCTOR INFORMATION

- Note how many conductors.
- Identify as copper or aluminum.
- Note wire size.

Photo #2 shows incoming service conductors in a main panel. Notice that both hot wires are black in this case. The neutral is taped white. You can identify the wires as copper or aluminum where they connect to the lugs. These are copper. By using your sample wire sizes, you can determine what size they are. Note also that 3 incoming service conductors indicates a 120/240-volt service.

Caution for Beginning Inspectors

If you're going to locate and examine your own main disconnect, please read pages 36 and 37 on safety precautions before opening the service box or main panel.

REMINDER: The home inspector should not determine amperage rating on the basis of service conductor size alone. Amperage rating is a little more complicated than that. Several items are checked for compatibility of amperage rating. See pages 28 and 29 for a discussion of this.

As the power enters the house, the service conductors go into a service box or the main panel. The hot wires will be red and black (or black and black) and are connected to the main fuses or circuit breakers. The third neutral or white wire does not connect to a fuse; it's connected to a neutral busbar.

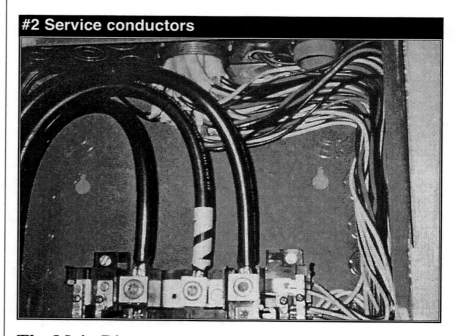

#2 Service conductors

The Main Disconnect

The incoming electrical service should have a main disconnect and main overload protection devices (fuses or breakers). This allows the power to be turned off to the house in a single movement.

Not every home has a main disconnect. A service without a main disconnect is not unsafe, although the NEC requires no more than 6 switches to disconnect all power to the house. These switches must be close enough together for all to be operated with no more than 6 hand movements.

The main disconnect may be located in one of the following places:

- In a service box outside at the meter (see drawing on page 19)
- In a separate service box inside
- Incorporated into the main panel

If the service box is outside, it should be inspected for waterproofing, rust, and whether or not it's securely fastened or bent and broken. Often, the owner has put a lock on the main disconnect box. This is a good idea as long as the owner knows where the key is in an emergency. Kids love to turn off electrical power on Halloween, and the padlock keeps them away. Service boxes inside should also be inspected for rust and corrosion.

A service box may be sealed by the utility, especially if located before the meter, to prevent homeowners from stealing power from the main disconnect. Mention a broken seal in your inspection report.

The home inspector can reaffirm amperage size by noting the size of the fuses or breakers at the main disconnect. The service box will have an amperage rating on it. Again, the inspector should not make a determination of amperage size by these features alone.

The main disconnect may consist of one or more of the following types of housings and overload protection devices:

- **Knife switches** that are turned off by means of a lever-type handle and protected by **cartridge fuses**. Please note that where there are 2 100-amp fuses in the service box as shown here, the house has a 100-amp service (if other service conductors and other items are compatible). Don't add the 2 fuse ratings together to get the amperage.

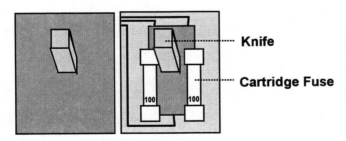

Knife

Cartridge Fuse

Service Box Covered and Open

- **Pull-out fuse blocks** that disconnect by means of a handle that permits pulling the block out of the box. With the pull-out block, the fuse size cannot be read without pulling the block and cutting power to the house. Before

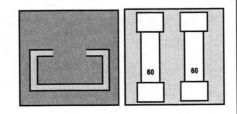

Pull-Out Block Front and Back

MAIN POWER

The home inspector does <u>not</u> have to turn off the main disconnect or test the main fuses or breakers. As a rule, do <u>not</u> turn off power during the inspection.

The only <u>exception</u> is pulling a fuse block to read the fuses. But be sure to ask the owner first if that will cause any problems.

pulling the block, check with the owner first to see if interrupting the power will cause any problems. Note too that the drawing of the pull-out block indicates a 60-amp service. You don't add the 2 fuses together.

There are usually 2 pull-out fuse blocks together in a panel. Usually the one on the **left** is the main disconnect; the other is commonly for the kitchen stove. That is, if you pulled the one on the left, the one on the right would lose power too. However, in some cases, **both blocks** have to be pulled to turn off the power completely. In that case, the fuses can be added between the 2 blocks to get the total amperage. For example, the 2 60-amps fuses in the left block give you 60 amps and the 2 40-amp fuses in the right block give you another 40 amps — a total of 100 amps.\

*Photo #3 shows a **pull-out block 60-amp service**. This is a typical setup with 2 blocks and 4 fuses below. We asked the owner first if we could pull the main block on the left, since all power to the house would turn off, and got permission first. We found 60-amp fuses in it. The right block will normally have 40-amp fuses for the kitchen stove. We checked that too. We did find one occasion where the blocks were switched and the right block was the main. In that case, we suggested that the owner have an electrician come in and switch them.*

#3 Pull-out block 60-amp service

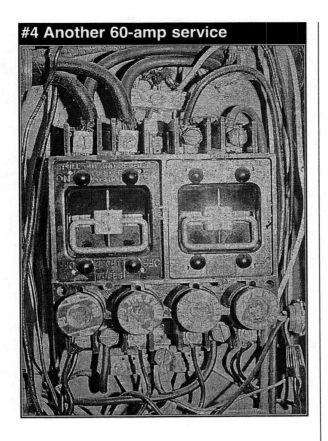

#4 Another 60-amp service

*Photo #4 shows another **60-amp service**. Notice the service conductors connected to the fuse block. Also notice that there are 2 more conductors connected to the fuse block. They lead to a subpanel. This is called **tapping before the main** and must be reported as a safety hazard. If you pull the main block, thinking you're turning off power to the house, those other 2 wires will remain hot and part of the house will still have power. We recommended that an electrician come in to redo this setup.*

- **Circuit breakers** are turned off like a wall switch. The main disconnect may be 2 or more breakers connected with a handle so they can be turned off at the same time. In the drawing shown here, these 200-amp breakers are connected by a handle between them. The service amperage rating would be 200 amps. You do not add them together.

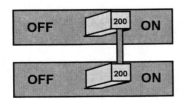

Photo #5 shows a **circuit breaker main disconnect**. Notice the main breakers connected by a handle. Each breaker reads 100 amps, indicating a 100-amp service. Again, you see **tapping before the main.** In this case, the 2 wires at the back are the service conductors. The 2 wires in front are connected at the main breaker and lead to a subpanel. This is a safety hazard and was reported as such. We recommended an electrician be called in to remedy the situation. There's another safety hazard in this photo. Look carefully. Notice that the conductors are both **burned** *(the grayish parts)*. With double tapping like you see here, arcing and overheating can result from a loose connection, burning back the wires. This is not a good situation.

#5 Circuit breaker main disconnect

Photo #6 is a circuit breaker **panel without a single main disconnect**. In this case, you would have to flick off the top 5 breakers to turn all power off to the house. The fifth breaker was for the main lighting for all the breakers underneath it. When there is no main disconnect, you should count the number of hand movements it will take to turn off the service. Any more than 6 movements is not allowed by the NEC.

#6 Panel without a single main disconnect

- **Screw-in fuses:** Only the oldest electrical systems around have screw-in fuses as the sole means of disconnecting the service.

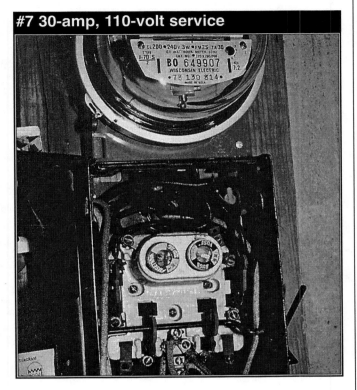

#7 30-amp, 110-volt service

*Photo #7 shows an old **30-amp, 110-volt service**. There are only 2 fuses servicing the whole house. The banks will no longer finance a home with this antiquated electrical service. We told the customer that it needed to be replaced.*

NOTE ON FUSES: **Type D fuses** are time delay fuses that do not blow immediately. They'll allow more than the rated current to flow through them for a short time before blowing. A **Type P fuse** has an added safety feature in that it is also sensitive to heat buildup between the fuse itself and the fuse holder. **Type S and C fuses** are non-interchangeable, and the wrong size fuse cannot be fit into the fuse holder.

When inspecting for the main disconnect, the home inspector should be sure to note these important problems:

- **More than 6 hand movements:** In cases where there is no main disconnect but many, count them to be sure the power can be turned off to the home with 6 or fewer hand movements. Report more than 6 as a **safety hazard**.

- **Incompatibility of components:** The home inspector may find the service conductors rated at a smaller amperage than the main fuses or breakers. This is not allowed and should be reported as a **safety hazard**. Allowing too much current to flow through underrated wires might cause excessive heat and possibly a fire. The service box may be rated at a smaller amperage than the conductors, fuses, or breakers. The box should be replaced with a compatible one.

- **Tapping before the main:** If the conductors are spliced before the main disconnect or there is tapping before the main, whatever is running off that line will continue to be live if the main disconnect is pulled. All power would not be turned off to the house. This too is a dangerous situation and should be reported.

- **Overheating, arcing, and burned wiring:** The home inspector should note the condition of the service conductors coming to the main breaker and inspect them for any evidence of overheating and arcing. Loose connections at this point can be dangerous, especially if there is tapping before the main disconnect.

Determining Amperage Rating

We finally come to the discussion of how to determine the amperage size of the electrical service. We've pointed out that the home inspector cannot rely on any *one* component of the service entrance. The determination is based on several factors.

The home inspector is **required to report the service amperage**. Take your time and examine several components of the service entrance system before coming to a conclusion. First, determine the amperage rating of the service conductors, the main fuses or breakers, and the service box or main panel. Check them against the rating of the meter and meter base. If they match, report that number as the service amperage. If they don't match, use the **lesser** or **weaker** of them to report the service amperage.

Don't report a 100-amp service if some components are rated at 100 amps and one component is rated at 60 amps. If you do and you're wrong, you'll pay to have the service upgraded to 100 amps. It's better to report it at 60 amps, then if you're wrong, it's good news for the customer.

Determine the service amperage by using the **lesser** of the ratings for the following:

- The size of the **service conductors** and their amperage rating. For example, if the service conductor is a #4 copper, then it is appropriate for a 100-amp service. A #2 aluminum conductor is also rated for a 100-amp service. (See chart on page 21.)

- The amperage of the **main overload protection devices** (fuses or breakers). For example, if the main disconnecting breakers are rated at 100 amps each, that would indicate a 100-amp service. Remember not to add fuse or breaker amperages together at the main disconnect.

- The amperage rating of the **service box or main panel** that houses the main disconnect. Somewhere on the panel should be a label stating its amperage rating. For example, the service box may be rated for 100 amps.

- The rating of the **meter and meter base**. As described on page 20, the home inspector can get a clue from the shape of the meter base and designations on the face of the meter what the service amperage may be. For example, a square meter base would be compatible with a 100-amp service. A CL200 designation on the meter face would be compatible for any service amperage up to 200 amps.

When the components listed above don't match, the home inspector should recommend that an electrician come in to fix the situation. A panel box which is rated at a smaller amperage than the main fuses or breakers should be replaced. Main fuses or breakers at a larger amperage than the service conductors should be replaced with the correct sizes for safety. Of least concern is the meter itself. Sometimes, the utility company may use an <u>old meter</u> in a <u>new meter base</u> when upgrading a service.

Grounding System

Grounding of the electrical system is required as a means of disposing of unwanted electricity. Grounding means connecting the electrical system to the earth. Before 1960, only the service panel required grounding, but since then all branch circuits, lights, and electrical outlets require grounding. Live or current-carrying conductors are not grounded; the neutral wire is.

SERVICE AMPERAGE

Choose the <u>lesser</u> of the amperage ratings of:

- Service conductors
- Main fuses or breakers
- Service box or main panel

For Beginning Inspectors

Locate the main disconnect in your home. Is it located in an outside or inside service box? Is it located in the main panel? (Read pages 36 and 37 on safety precautions before opening any boxes or panels.)

Use the sample wires you purchased and try to identify the gauge of your service entrance conductors. What is their amperage rating? Identify the type of main disconnect if there's one present. What is the amperage rating of the main fuses or breakers? What does that indicate about the service amperage rating? Find an amperage rating on the service box or main panel. Take a look at your meter and meter base. What is its amperage rating?

Finally, what is the service amperage rating for your house?

Grounding is the process of electrically connecting any electrically conductive item to the earth.

Bonding means electrically connecting 2 or more conductive items together and to the grounding system.

Bonding means electrically connecting all conductive items in the electrical system together by means of a conductor to the grounding system. Conductive items include all boxes, panels and subpanels, conduit, junction boxes, cable armor, appliance cabinets, metal bathtubs and sinks, and so on. They must all be bonded to each other and to the ground.

At the main panel, the incoming neutral, or white, service conductor is connected to a **neutral busbar** and typically to the panel itself. The ground wire from each individual branch circuit also terminates at the neutral busbar. A conductor bonds the main panel itself to the busbar. The electrical system is grounded by means of a **grounding conductor** which runs from this busbar to the plumbing system, house footing and/or to a rod or rods driven into the earth.

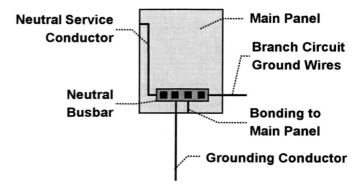

The grounding system could be at the meter enclosure or at the main disconnect located in a separate service box. If that's the case, there should be a ground conductor between them and the main panel (in addition to the 2 hot conductors and the neutral). The grounding conductor should not be spliced.

For grounding purposes, a #8 copper conductor can be used for up to 125 amps, #6 for up to 175 amps, and #4 for up to 200 amps. But when grounded to a driven rod, #6 copper is the largest required. Aluminum would require a larger size in each case. Aluminum will corrode when exposed to moisture, so when it is used as a grounding conductor, its connection to the ground rod should be at least 18" above grade.

The grounding conductor may be connected to the plumbing system near its point of entry in the house. If it is connected on the house side of the water meter, a jumper wire should be provided across the water meter. Of course, the grounding conductor may not be connected to plastic water pipes — metal only. Since 1987, the NEC has also required a copper or stainless steel rod driven 8' into the earth in addition to the plumbing ground. The latest NEC ruling may require 2 driven ground rods, depending on resistance. So, older installations may have a driven ground, a water ground, or both.

The actual inspection of the grounding system is fairly minimal. The home inspector is not required to determine the capacity of the grounding system. It is, however, important to be able to say if the system is grounded and that what's visible looks to be proper. Check for the following:

- **Missing ground:** Look for the presence of an appropriately connected grounding conductor in the main panel. Sometimes when a new panel box is installed, the grounding conductor may not have been moved from the original box. Recommend that an electrician come in to put in a missing grounding system or to reconnect one.

  It may be that you simply cannot locate the grounding system and you don't know if it's truly missing. In that case, you should inform your customer and tell them to have an electrician be called to verify whether or not a grounding system exists.

- **Splices in grounding conductor:** The grounding conductor requires special tools to splice. A splice is a potentially weak connection.

- **Lack of bonding:** As discussed on page 30, the neutral conductor, the branch circuit ground wires, the grounding conductor, and the main panel itself should be bonded at a neutral busbar. Note any missing bonding or defects with the busbar. An electrician should be recommended to fix this situation.

- **Ineffective grounding:** Check the clamp that connects the grounding conductor to a water pipe or rod to be sure it's not loose, broken, rusted or corroded so as to break the electrical connection. Follow the plumbing pipe after the grounding connection to see if any portions have been replaced with plastic and no longer provide a ground. Notice also if the electrical system is grounded to old, no longer used water pipes. Sometimes, new piping is installed but the ground is not moved to the new pipes. It should be.

INSPECTING GROUNDING SYSTEM

- Missing ground
- Spliced grounding conductor
- Lack of bonding
- Loose, broken, rusted, or corroded clamps
- Plastic plumbing or dielectric pipe connectors

Definition

A dielectric connector is used between 2 dissimilar metal pipes that prevents metal-to-metal contact between the pipes and stops the flow of electricity along the pipes.

For Beginning Inspectors

See if you can locate the grounding system in your own home. Notice if any of the problems we've discussed are present.

Another problem that can appear with metal water pipes is the use of a **dielectric connector** between 2 different metal water pipes. This type of connector removes the metal-to-metal contact between lengths of pipe. The grounding conductor should be connected after this type of connection.

Reporting Your Findings

As you begin the inspection of the service entrance, as we've described in these pages, have your customer come with you. But, as cautioned, don't allow the customer to get too close to what you're inspecting. Always watch out for your customer's safety when it comes to the electrical service.

One important aspect of the home inspection is the **education** of the customer about the home. Nowhere is that job as important as when you are inspecting the electrical system. So few people know how electrical systems work. Actually, few people *want* to know about it. The home inspector, however, is still responsible for educating the customer about things they *need* to know.

The home inspector walks a fine line. You must communicate your findings, and you must also put those findings into words that will reach and educate the customer. And some people have a high resistance. We've had cases where we pointed out electrical safety hazards only to have new homeowners ignore the warnings and lose their home to fire. So use simple terms and work with customers to **help them understand what you're telling them**. Be patient. Explain things clearly without giving a boring technical lecture. And take the time to answer any questions the customer may have.

As you are performing the inspection of the service entrance components, explain the following patiently:

- **What you're inspecting** — the service drop, the exterior conduit, the service entrance conductors, the main disconnect, the grounding system.

- **What you're looking for** — frayed wires, loose masthead, overheating or arcing, spliced grounding conductors, and so on.

- **What you're doing** — comparing wire gauges to your samples, counting hand movements required to turn off the power, following the water pipe to see if any plastic sections have been added, and so on.

- **What you're finding** — improper clearance over driveway, broken meter seal, tapping before the main disconnect, improper fuse sizes, and so on.

- **Suggestions about dealing with the findings** — calling the utility company to fix frayed overhead wires, having an electrician come in to fix ineffective grounding, and so on. But with this caution — don't make uneducated guesses about how repairs should be made.

Filling in Your Report

Every home inspector needs an inspection report. A **written report** is the work product of the home inspection, and every home inspector is expected to deliver one to the customer after the inspection. Inspection reports vary a great deal in the industry with each home inspection company developing its own version. Some are considered to be excellent, while others are not very good at all. A workable and easy to use inspection report is very important for a home inspector in terms of being able to fill it in. Of greater importance is its thoroughness, accuracy, and helpfulness to your customer. We can't tell you what type of inspection report to use, but let's hope that it's a professional one.

The **Don't Ever Miss list** on the opposite page is a reminder of those specific findings you should be sure to include in your inspection report. We list these items after years of experience performing home inspections. Missing them can result in complaint calls and lawsuits later. Here is an overview of what to report on during the inspection of the service entrance:

- **Exterior service drop:** Report whether the service is overhead or underground and whether it's adequately supported. Report any deficiencies such as improper clearances and tapping before the meter, for example. Make note of whether you advise the customer to call the utility company to remedy any situations.

- **Main disconnect:** Identify the type (copper, aluminum, etc.) of service conductors present. Be sure to report what you've determined to be the service's amperage and voltage ratings. Note whether fuses or breakers are present. Report deficiencies such as an inadequate 30-amp

Report Available

The American Home Inspectors Training Institute offers both manual and computerized reports. These reports include an inspection agreement, complete reporting pages, and helpful customer information.

If you're interested in purchasing the Home Inspection Report, please contact us at 1-800-441-9411

DON'T EVER MISS
• Improper service drop clearances
• Unsupported mast, conduit or cable, or meter
• Tapping before meter or before main disconnect
• Oversize fuses or breakers for wire size
• System not grounded
• Lack of bonding
• More than 6 hand movements to turn off power
• And never guess at amperage size!

or 60-amp services that should be replaced and oversize fuses. Note whether you recommend an evaluation by a licensed electrician because of some condition present.

• **Grounding:** Report on whether the system is grounded or not. Report deficiencies on the grounding systems such as grounding conductors not connected to new water pipes, for example.

• **Safety hazards:** Never miss reporting any safety hazards you've found. It's a good idea to report them on the page of your report that specifically deals with the subject and then summarize them on a summary page at the back of the report. For the inspection of the service entrance, don't miss these safety hazards:

— Masthead within reach of occupants
— Frayed cable on house exterior
— Tapping before the main disconnect
— Oversize fuses or breakers in main disconnect
— System not grounded
— More than 6 hand movements to turn off power

WORKSHEET

Test yourself on the following questions.
Answers appear on page 36.

1. How many feet of clearance to the overhead service drop does the NEC require over driveways?

   A. 8'
   B. 10'
   C. 12'
   D. 18'

2. Which of the following indicates a 120-volt electrical service?

   A. 1 conductor entering the masthead and 2 wires tied back
   B. 2 conductors entering the masthead and 1 conductor tied back
   C. 3 conductors entering the masthead
   D. 4 conductors entering the masthead

3. Which shape of meter base may be an indication that the home has a 60-amp electrical service?

   A. Round
   B. Square
   C. Triangular
   D. Rectangular

4. What wire gauge should the service conductors be for a 100-amp electrical service?

   A. #6 copper or #6 aluminum
   B. #4 copper or #2 aluminum
   C. #2 copper or #4 aluminum
   D. #2/0 copper or #4/0 aluminum

5. A main disconnect with 2 60-amp fuses would indicate what amperage rating?

   A. 30 amps
   B. 60 amps
   C. 100 amps
   D. 120 amps

6. Which of the following components is <u>not</u> considered when determining the service amperage rating?

   A. Service conductor amperage rating
   B. Main panel amperage rating
   C. Main fuse or breaker amperage
   D. Exterior conduit size
   E. Shape of the meter base

7. What is considered to be a safety hazard at the main disconnect?

   A. The use of cartridge fuses
   B. 5 hand movements to turn off power
   C. 2 main breakers connected by a handle
   D. Extra wires tapped before the main

8. Which statement is <u>false</u>?

   A. The home inspector must describe the amperage and voltage rating.
   B. The home inspector must identify the service conductor materials.
   C. The home inspector must turn off power to the house.
   D. The home inspector must report whether the service drop is underground or overhead.

9. Which of the following grounding methods is <u>not</u> acceptable to the NEC?

   A. Grounding to gas pipes
   B. Grounding to water pipes
   C. Grounding to a driven ground rod or rods
   D. Grounding to water pipes and a driven rod

10. Some electrical services don't need to be grounded.

    A. True
    B. False

Guide Note

Pages 36 to 51 present procedures on inspecting the main panel and subpanels.

Worksheet Answers (page 35)
1. *C*
2. *B*
3. *A*
4. *B*
5. *B*
6. *D*
7. *D*
8. *C*
9. *A*
10. *B*

Chapter Four

INSPECTING THE MAIN PANEL

It's at the main panel and subpanels that the home inspector gets a good idea of the condition of the electrical system. A professional job here can be a reassurance that the system is done properly. A sloppy job at the main panel should set off alarm bells in the inspector's mind.

Safety Precautions

Home inspectors should think about their customer's safety and their own safety before approaching the main panel. Here are the steps to take before you remove the cover to the panel and begin to inspect it:

- **Have your customer stand back,** preferably a few feet behind you and to the side. If you get a shock from the main panel and get thrown back, you won't crash into the customer. Never let the customer approach the main panel first or right at your side.

- **Note the presence of water.** If there's standing water on the floor beneath the panel, do not stand in it to inspect the main panel. If there's so much water that you can't get close to the panel, don't inspect it. Make a note in your inspection report that the panel was inaccessible and not inspected. If you can see water from above leaking into the panel, do not inspect the panel. Report that the panel was not inspected due to water in the main panel.

- **Listen for arcing.** If you can hear the buzz of arcing going on in the panel, do not remove the panel cover and inspect the panel. Report the condition and explain that it was unsafe to inspect the panel.

- **Test for current.** First, put your rings in your pocket before you approach the main panel. And leave your screwdriver in your pocket until you've tested the box for current. Put the back of your right hand, your knuckles, against the main panel. If it's live with current, you'll receive a shock and your hand will fly back against your body. You may hit yourself in the face that way, but at least your arm won't be wrenched outward as it would if you touched the panel with the palm of your hand. It's a good idea to develop the practice of keeping your left hand behind you when you touch the panel the first time. If your right hand receives a shock, the current will pass down the right side of your body and not the left where the heart lies.

If you do receive a shock, do not approach the main panel again. Leave it uninspected. Write it up in your inspection report as unsafe to inspect because of live current at the main panel.

- **Feel the box for heat.** After you're sure there's no live current flowing in the panel box itself, touch it again and feel if it is hot or warm to the touch. A warm panel means that wires inside are overheating due to loose connections. Such a panel is dangerous. Again, do not open or inspect the panel, but be sure to report why you didn't inspect it.

Panel Locations

The home inspector should note the **location** of all panels and subpanels in the inspection report and should check for the proper clearance around the panels.

The NEC requires any panels to be located at least 5 1/2' above the floor with 6' 3" of headroom. The top of the panel should not be over 6 1/2' high. The wall below the panel should be clear to the floor. The panel should be in a clear space at least 30" wide with at least 3' of unobstructed space in front of the panel. Panels cannot be located in enclosures such as closets or placed near flammable liquids. However, you'll often find that many multi-family homes or condos have panels in closets. Check with your local inspector for local requirements.

#8 Dangerous location for a main panel

Photo #8 shows a *dangerous location for a main panel.* Yes, that white piping is a showerhead. That's a bit close for comfort — an obvious safety hazard. In this case, the shower was still hooked up and even if the customer doesn't plan on using the shower, it should be taken out. By the way, this is a different style of panel. The main disconnect at the top is a single block. Below you can see a series of fuses around 2 more 240-volt blocks.

Watch for main panels and subpanels located in unsafe places.

MAIN PANELS

- Single-bus panel
- Split-bus panel

Main Panel Layouts

The **main panel** is a metal box that carries overload protection devices and/or disconnects for circuits for the home's electrical service. The main panel has **busbars** or conductor bars that provide connections for fuses or breakers. These connections are lined up like seats on a bus, hence the name.

The **single-bus panel** has a single pair of busbars that provide power to fuses or breakers for either 120-volt or 240-volt circuits. The live or hot service conductors are usually connected to a **main disconnect**, either a fuse block or main breaker, and 120

Definitions

The <u>main panel</u> is a metal box that carries overload protection devices and/or disconnects for household circuits.

A <u>busbar</u> is a conductor bar that provides connections and power for fuses or breakers.

A <u>single-bus panel</u> is a panel with a single pair of busbars.

SINGLE-BUS PANEL

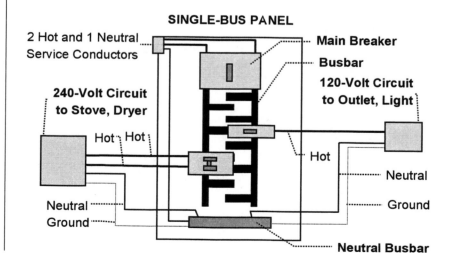

volts is provided for each leg of the busbar. Breakers for 120-volt circuits (1-pole breakers) and breakers for 240-volt circuits (2-pole breakers) can be arranged anywhere on the busbar. 120-volt circuits use only one leg of the busbar, while 240-volt circuits use both legs.

The **120-volt circuit** is provided power between one hot wire and the neutral wire, which is connected to a neutral busbar within the panel. (See page 12 for an explanation of household circuits and page 29 for grounding procedures). The **240-volt circuit** is provided power between 2 hot wires, and in some cases, the neutral wire may also run to the appliance on the circuit. Both types of circuits have a **ground wire** to bond the outlet to the grounding system.

The 2 breakers (or fuses) used for 240-volt circuits should be the same size and linked together so that if one is pulled or turned off, the other must also be pulled or turned off. There should be a handle connecting 2 individual breakers or a single 2-pole breaker should be used.

Take another look at **Photo #5** in this book. This is a single-bus panel with the main disconnect at the top. Notice the overload devices on the left for 240-volt circuits — a single handle turn-off and 2 hot wires from each one. The devices for 120-volt circuits are on the right, each with a single hot wire emerging. Remember, we stated that double tapping off the main disconnect was a safety hazard and not allowed. Here, you can see the wires at that connection burned back.

A **split-bus panel** has 2 or more pairs of busbars that provide power to fuses or breakers. In split-bus panels there is no main disconnect. The **upper busbar** usually provides room for up to 6 fuses or breakers for 240-volt circuits — 5 for major appliances and a sixth as the main overload device for the lower busbar. Note that this complies with NEC requirements to be able to turn off power with 6 hand movements.

The **lower busbar**, fed by the 2-pole breaker from above, provides power for all the 120-volt circuits in the house. This 2-pole breaker may be labeled *Lighting* or *Main Lighting*. Turning off this breaker will turn off the power to all the 120-volt circuits in the lower busbar.

You may find another 2-pole breaker on the upper busbar that

Definitions

A split-bus panel is a panel with 2 or more pairs of busbars.

The upper busbar usually provides connections for up to 6 240-volt circuits, one of which provides power to another busbar. The lower busbar provides connections for all the 120-volt circuits in the house.

energizes a third busbar of 120-volt circuits.

SPLIT-BUS PANEL

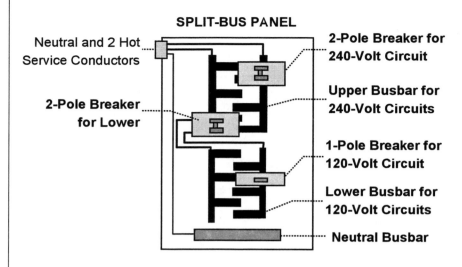

Neutral and 2 Hot Service Conductors

2-Pole Breaker for 240-Volt Circuit

2-Pole Breaker for Lower

Upper Busbar for 240-Volt Circuits

1-Pole Breaker for 120-Volt Circuit

Lower Busbar for 120-Volt Circuits

Neutral Busbar

#9 Split-bus panel

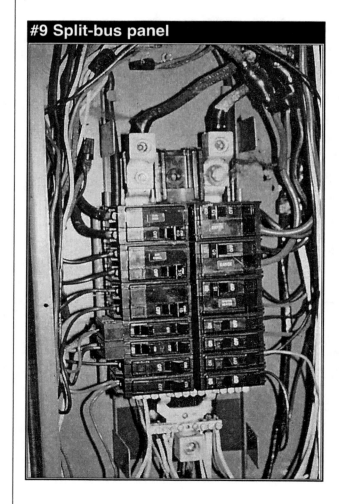

Photo #9 shows a *split-bus panel*. Note first that there is no main disconnect, and the 2 hot service conductors are connected to lugs at the top of the panel. The top 6 breakers are for 240-volt circuits — 2-pole breakers, each with a single turn-off switch, and 2 hot wires emerging from each one. The bottom 8 breakers are for 120-volt circuits. Notice too the neutral busbar at the bottom with all the white ground wires for each circuit in the house. The lug centered below the neutral busbar is the connection for the neutral service conductor (thicker black wire).

Subpanels

Electrical systems can be upgraded over time. When more power is added to the system, there may be additional main panels added or the main panel may be upgraded. In this case, power is actually increased to the house and new circuits are added to distribute it. There may be a trough added that feeds power to both main panels. When there is more than 1 main panel, there still must be only 6 hand movements to turn off the power to the house and all disconnects should be within reach of each other.

Sometimes other panels are added to existing systems not for the purpose of increasing power, but merely to provide more circuits and better distribution. In that case, the additional panels or boxes are called **subpanels**. The home inspector is required to remove the covers of the subpanels, if safe, and inspect the conditions inside them.

The home inspector may find the following wiring techniques between main panel and subpanel:

- One of the **240-volt circuit breakers** in the main panel supplies power to the circuits in the subpanel. It's like a third busbar at a remote location. This is an acceptable practice if wire gauges and fuse or breaker sizes are both rated correctly. For example, #4 copper or #2 aluminum wire would be needed for a 100-amp fuse or breaker running to the subpanel (see chart on page 21).

- **Special lugs** may be provided in the main panel to which feeder wires are attached that feed power to a main overload device in the subpanel.

- **Tapping before the main disconnect** in the main panel to feed the subpanel is **not allowed**, although the home inspector may see examples of this. If the main disconnect in the main panel was turned off, the subpanel circuits would still be live. This is a dangerous situation and should be reported as a **safety hazard**.

SUBPANEL WIRING

- Feed from one of the 240-volt breakers in the main panel.

- Feed from special lugs in the main panel.

- Not allowed: Tapping off the main disconnect in the main panel.

- The neutral wire must be isolated for the subpanel. (Wire cannot come into contact with the panel.)

- The neutral wire (white) and the grounding wire must be on separate grounding bars.

- The grounding wire must be in contact with the panel.

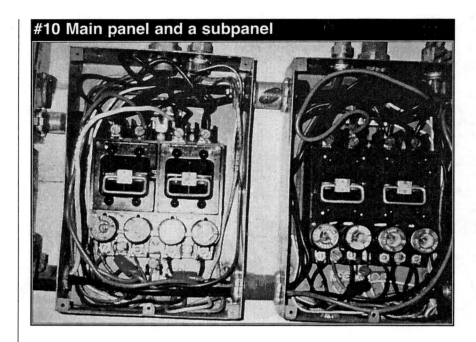

#10 Main panel and a subpanel

Photo #10 in this book shows a *main panel and a subpanel*. If you look closely, the main disconnect in the main panel (fuse block on the left) is tapped before the main and those are the wires feeding the box on the right. This is illegal wiring and should be reported as such. .

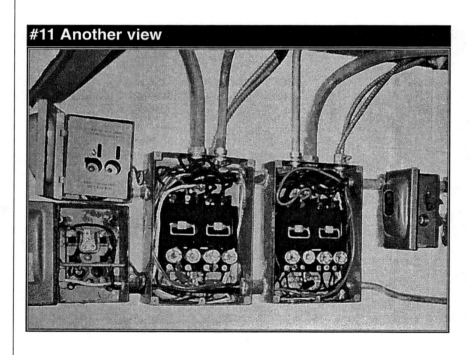

#11 Another view

Photo #11 shows **another view** of these same boxes. Here, you can see that more boxes have been added downstream of the second panel. You can't run wires through one box to another to get to a third box

NOTE: The subpanel should be bonded to the grounding system in the main panel box. But within the subpanel, circuit ground wires and the neutral wire should not be bonded to each other in the subpanel.

Inspecting the Panels

The home inspector should inspect each main panel and subpanel present, if safe and possible. Watch for the following conditions:

- **Loose, rusted, wet, or uncovered panels:** The home inspector should inspect each panel present. Follow the safety precautions as outlined on pages 36 and 37 when approaching each panel. Note if there is a cover present on the panel. Often times, covers are missing and panels are left open with the wiring exposed. This should be noted in your inspection report as a **safety hazard**.

  Check the general condition of each panel itself. Is it secured to the wall? Are there any signs of rusting? Is there water leaking into the panel? Are there proper clearances? Report these conditions in the inspection report.

- **Undersized panel:** Locate the rating label on all boxes and confirm that each box is rated at the proper amperage for the service. Any main panel or subpanel should be replaced if its amperage rating is lower than the determined service amperage.

- **Missing grounding conductor:** Look for the presence of a grounding conductor in the main panel. See pages 29 to 31 for the inspection of the grounding system.

- **Undersized main disconnect:** The fuses or breakers at the main disconnect must be compatible with the service amperage rating. Refer to pages 22 to 28 for information.

- **Damage where service conductors enter:** Look at the fittings at the back or side of the box where the service conductors enter. The fittings should protect the conductors from excessive wear.

- **Tapping before the main:** Again, notice the connections of the service conductors to the main disconnect. Are other wires tapped to the main lugs? Report the situation as a **safety hazard**.

INSPECTING PANELS

- Loose, rusted or wet, uncovered panel boxes
- Undersized panel
- Missing grounding conductor
- Undersized main disconnect
- Wear at entry fittings
- Tapping before the main
- Arcing, burned wiring, and melted insulation

Guide Note

At the main panel, the home inspector would also be inspecting the main disconnect and the grounding system. See pages 22 to 32 for a review of that portion of the inspection.

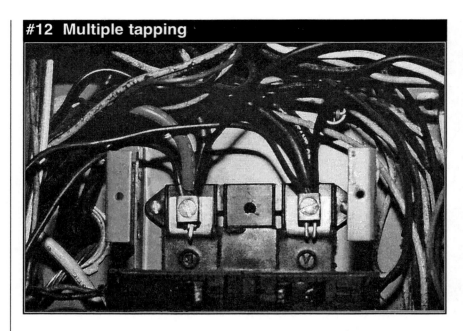

#12 Multiple tapping

Photo #12 shows *multiple tapping* at the main lugs. There are at least 4 wires at each lug. You wonder sometimes how far people will push the equipment. The home needed more circuits, and someone had continued to add them off the main disconnect. An electrician should be called in to redo the wiring and add some subpanels using acceptable methods.

- **Burned wiring and melted insulation:** Watch for overheating at the main disconnect, burned back wires, and melted insulation.

#13 Aluminum service conductor

Photo #13 shows an *aluminum service conductor* at the main that was arcing. The heat had been so intense that the insulation has burned way back. An electrician must be called to fix this problem.

- **More than 6 hand movements:** As you inspect the main panel and subpanels, be sure to count the number of hand movements required to turn off the power. More than 6 turnoffs should be written up in your inspection report as a problem.

#14 Multiple panels

*Photo #14 shows **multiple panels** along a basement wall. This was a duplex where we found around a dozen different boxes with wires going everywhere. There were 10 different turn-offs, and they weren't close together. And notice that 3 of the larger panels are missing covers and have the wiring exposed. This system just isn't up to standard. We recommended that a professional electrician be called in to clean it up.*

Inspecting the Fuses or Breakers

The next step in inspecting the main panel and subpanels is to look at the rest of the fuses and breakers in the box. The home inspector does *not* have to test fuses and breakers, except to test for the operation of GFCI's. And remember, the inspection of the box is a visual inspection. You don't need to stick your fingers into the panel box to notice the following defects:

- **Overfusing:** With some types of fuses, it's possible to replace the proper size fuse with another size. The typical household circuit is a 15-amp circuit requiring a 15-amp fuse and #14 copper or #12 aluminum wire. A homeowner can put a 20-amp fuse into the fuse holder, causing the wire to overheat and create a safety hazard. Appliances such as the clothes dryer may need a 30-amp fuse and larger wiring; the kitchen stove may require a 40 or 50-amp fuse and compatible wiring. Check the fuse sizes against the wire sizes (see charts on pages 13 and 14).

Inform the customer that oversized fuses should be replaced. There are type-S adapters that can be purchased to fit into the fuse holder that will reject any wrong-sized fuses.

INSPECTING FUSES AND BREAKERS

- Overfusing
- Scorched cartridge fuses
- Broken or cracked fuse holders
- Pennies or foil present
- Open knockouts
- Double tapping
- Inoperable GFCI's
- Non-approved breakers for aluminum branch wiring
- Unlinked breakers

"Just when you think you've seen everything, something new comes along.

"One of my inspectors had removed a panel cover not noticing that there was an odd screw among the original ones. When he was replacing the cover, he screwed in the odd screw. The screw was too long and as he tightened it, the screw hit the main lug. Well, the screw shot out of the panel like a bullet and zoomed all the way across the basement, hitting the far wall.

"I still shudder to think what could have happened if the inspector or the customer had been in the path of the screw."

Roy Newcomer

Definition

A GFCI is a ground fault circuit interrupter. It's a monitoring device that will trip in 1/40 second after a ground fault of only .005 amps is detected, interrupting the flow of electricity to the circuit.

- **Scorched cartridge fuses:** If cartridge fuses have scorch marks on them, it's an indication of long-term overheating. Suggest that an electrician be called in when you find any signs of overheating in the panel.

- **Broken, cracked fuse holders:** In an old box, the porcelain holders can crack or break, causing poor connections, loose fittings, and corrosion. If you see this condition, suspect overheating too.

- **Pennies or foil present:** People sometimes wrap blown fuses with foil or insert a penny in a fuse block. These are dangerous practices and a safety hazard. Suggest that fuses be replaced and any foil and pennies be removed.

- **Unused knockouts:** If a circuit is no longer used for some reason, the fuse holder or knockout may remain empty. Any open holes should be sealed as there is still power flowing to the area and touching it could cause a severe shock.

- **Double tapping (120 volts):** People will often add more circuits to the house by tapping another wire at the fuse or breaker lug (or screw). With breakers, it is allowed only if the breaker is approved for double tapping. But in general, only one wire should be present at the fuse or breaker lug. When double tapping is present, it's an indication that a subpanel may need to be added for the additional circuits. It's best to check with your local electrical inspector on how they handle this. Our practice is to write in the report that double tapping is not recommended. Double tapping of 240-volt circuits is written up as a **safety hazard**.

- **Inoperable GFCI's:** The home inspector should note the presence of **ground fault circuit interrupters** breakers in the panel boxes and test them for operation. Although most GFCI's are for circuits near water, outside, and in the garage, it's a good idea to check with the owner just to be sure before interrupting the power on these circuits.

To test the GFCI, press the test button. The first test is whether the breaker will switch off. Next, turn the breaker back on.

The second test is if the breaker will go back on. If the breaker won't engage, the GFCI breaker is defective. If it's defective, you have now turned off power to that circuit, so be sure to tell the owner or leave a note for the owner about the defective GFCI. *You don't use a GFCI tester at the main panel.*

- **Non-approved breakers for aluminum wiring:** We're going to be talking a lot more about aluminum branch circuit wiring starting on page 62. For now, note that if the branch wiring is aluminum, then the breakers must be marked **AL** (aluminum) or **CO/ALR** (copper, aluminum revised). Any other breaker type is a safety hazard.

- **Unlinked breakers:** Where 240-volt circuits are protected by 2 breakers, those breakers should be linked with a handle so that turning off one breaker turns off the other.

Inspecting the Wiring

The final step in inspecting the panel boxes is to examine the **branch wiring** — the wiring at the fuses or breakers for the household circuits. (Pages 53 to 64 will present more detail on branch wiring.) At the boxes, identify the type of wire as aluminum or copper and then look for the following conditions:

- **Incompatible wiring:** Wiring from fuses and breakers should be the appropriate gauge for the amperage rating of the fuse or breaker. Allowing too much heat to flow through underrated wires might cause excessive heat and possibly a fire. This should be reported as a **safety hazard**.

- **Damaged wiring:** Make note of any nicks in the wiring in the panel box. Sometimes, wires can get pinched when the panel cover is put back on. A wire that is nicked cannot carry its rated amperage and can be overloaded.

- **Overheating, arcing, melted insulation:** Check the branch wiring connections to the fuses or breakers and report any evidence of overheating, arcing, and melted insulation. These conditions are usually the result of loose connections at the lugs. An electrician should come in to assess the situation and fix it.

INSPECTING WIRING

- Incompatible wiring
- Damaged wiring
- Overheating, arcing, and melted insulation
- Unprotected splices or abandoned wires
- Missing anti-oxidant on aluminum wiring
- Handyman wiring

An <u>anti-oxidant compound</u> is a grayish paste applied to aluminum wiring connections to prevent aluminum oxide from forming on the surface of the wire.

- **Unprotected splices, abandoned wires:** Splices of branch wiring within the panel is allowed when a longer length of wire is needed for a subpanel although the splices must be properly protected. However, splicing wires for 2 circuits so they can use the same breaker is not allowed. Any wires that are abandoned in the panel should be appropriately terminated or removed so there is no chance of them touching a live component.

- **Missing anti-oxidant on aluminum wiring:** All aluminum wiring connections, including the connection to the fuse or breaker lug, require an anti-oxidant to be applied. The **anti-oxidant or anti-corrosion compound** is a grayish paste. Without it, aluminum wiring will oxidize, and aluminum oxide is a good insulator. Therefore, oxidation will cause a poor connection and become overheated, perhaps enough to cause a fire.

  Look back at **Photo #13**. This **aluminum service conductor** has been seriously burned back due to overheating. Anti-oxidant should also be present where aluminum service conductors are connected to the main lug.

- **Handyman wiring:** Any evidence of handyman or amateur wiring at the main panel should put the home inspector on notice that all might not be right with the rest of the service. Whenever you see bizarre and obviously illegal wiring practices, let the customer know that an electrician should be called in to examine the entire electrical system in the house.

*Photo #15 shows **handyman wiring**. Obviously, the main panel isn't sufficient to service the house, so the owner has increased the number of circuits. There's tapping off the main fuse block and double tapping off the fuse lugs at the bottom. Notice the black cable projected outward and to the left. Wires flow from it directly outside the box. The panel cover could not even be put on. There's no solution for this but to call in an electrician and start over. And that's what we recommended in our inspection report.*

#15 Handyman wiring

#16 Inventive Handyman wiring

*Photo #16 shows another example of **inventive handyman wiring**. It's a little hard to tell what's happening with the main panel, but notice the wiring outside the panel. The owner has created another circuit, and even installed a fuse for it — outside the box! Sorry, but that just doesn't cut it. The panel cover can't be put back on the box to protect anyone from touching this mess. We recommended that an electrician be called to redo the work and check out the rest of the electrical system for other amateur wiring.*

Now that you've seen some examples, it should be crystal clear as to why we warn you not to touch things inside the main panel and subpanels. Many dangerous situations can be present. Always be careful when inspecting the panels.

Reporting Your Findings

Talk to your customer during the main panel inspection, but have them stand back and to the side, out of harm's way. Don't let your customer touch anything in the main panel. Consider yourself responsible for the customer's welfare.

Be sure to explain to your customer what you were inspecting at the main panel and subpanels and what you found. Take the time to answer questions. Remember that customers may not understand what they see at the main panel and are depending on you to make some sense out of it for them. Be sure to **stress safety hazards** you find and tell the customers that you will also indicate safety hazards in the inspection report for them. Suggest that customers review the inspection report again on their own.

Personal Note

"Mice can get into everything. I've opened a panel and found a mouse nest right in the box. That's not good. Mice can nibble at insulation and make a mess of the box. I'm not sure why more mice aren't electrocuted in the process."

Roy Newcomer

When reporting on your inspection of the main panel and subpanels, be sure to report on the following:

- **Panels:** Record the **location** of each one you find. Be sure to write **whether or not you were able to evaluate** the panel. For example, you might write "Panel not evaluated due to live current" or "Panel not evaluated due to water present on floor" or whatever the reason. Report defects such as improper clearances for the panel, rusted box, or underrated panel and safety hazards (see opposite page).

- **Fuses and breakers:** Report whether fuses and breakers are present and any deficiencies you find as noted in the Don't Ever Miss list. Also, indicate if you've found a GFCI in the panels and report whether it was operating or not.

- **Wiring:** The home inspector should identify the type of branch circuit wiring (copper, aluminum) in the panels and report on its condition at the panel connection. Don't miss multiple tapping, arcing, burned wiring, and melted insulation.

- **Recommending an electrician:** The home inspector should always suggest that the new homeowner have a professional electrician come in to evaluate the main panel and subpanel when serious conditions are found. This should be noted in the inspection report to help your customer remember your recommendation. Don't suggest an electrician when one isn't needed. However, the list of safety hazards below and the conditions listed here are cause to have an electrician involved:

 — Multiple tapping
 — Multiple boxes and too many turn-offs
 — Wires overheated, burned, damaged
 — Melted insulation
 — Live current in the panel
 — Double tapping off fuses or breakers
 — Handyman wiring

- **Safety hazards:** All safety hazards found during the inspection of the main panel and the subpanels should be noted in the inspection report.

    — Live current in main panel
    — Uncovered panel with wires exposed
    — Oversize fuses or breakers
    — Undersize wires for fuses or breakers
    — Supplemental wires tapped into main lug
    — Unlinked breakers
    — System not grounded
    — More than 6 hand movements to turn off power

NOTE: We stress the importance of accurate and detailed reporting because of the liability the home inspector has regarding the electrical system. We're sure you get the idea. If you miss reporting findings listed in the Don't Ever Miss list, you know you'll hear about it later in the form of a complaint call. The home inspector who misses these details in the inspection report will only have to pay for electrical work later.

DON'T EVER MISS

- Loose, rusted or wet, uncovered, and undersize panels
- Double tapping at the main lug or fuses or breakers
- Arcing, burned or damaged wiring, and melted insulation
- Incompatible fuses and wiring, overfusing
- Cracked fuse holders, open knockouts, abandoned wires
- Inoperable GFCI's
- Unlinked or non-approved breakers
- Handyman wiring

WORKSHEET

Test yourself on the following questions.
Answers appear on page 54.

1. Which of the following is the correct NEC requirement for panel clearance?

   A. There should be 5 1/2' of headroom.
   B. There should be 5 1/2' of clear space in front of any panels.
   C. The panel should be 5 1/2' from the floor.
   D. There should be 5 1/2' clearance at each side of the panel.

2. When should the home inspector <u>not</u> remove the cover to the main panel?

   A. If your customer is present
   B. If there are multiple panels
   C. If the meter is inside the building
   D. If you get a shock upon touching it

3. What type of main panel is shown here?

   A. Single-bus panel
   B. Split-bus panel

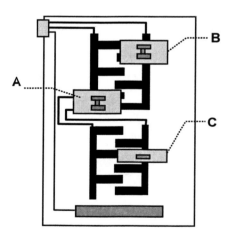

4. In the drawing above, which breaker supplies power to all the 120-volt circuits — A, B, or C?

5. In the drawing above, which breaker might be used for the circuit for the kitchen stove — A, B, or C?

6. Which of the photos in this guide shows proper wiring at the main lugs or main disconnect?

   A. Photo #5
   B. Photo #9
   C. Photo #12
   D. Photo #13

7. Which is an example of overfusing?

   A. Using a 20-amp fuse for a 15-amp circuit
   B. Tapping 2 circuits at a fuse lug
   C. Splicing another circuit onto a fused circuit
   D. A fuse that has foil wrapped around it

8. What is a subpanel?

   A. A second main panel in a duplex
   B. A panel that adds more power to the service
   C. A panel that adds more circuits to the service
   D. A panel housing the main disconnect for the service

9. Which of the following photos shows a split-bus panel?

   A. Photo #5
   B. Photo #6
   C. Photo #7
   D. Photo #9

10. When inspecting the fuses or breakers, the home inspector should:

    A. Touch the branch circuit wiring to feel for heat
    B. Jiggle the wiring to see if it's secure
    C. Compare amperage rating of fuse or breaker and branch wiring
    D. Test the fuse or breaker for operation

Chapter Five

INSPECTING BRANCH CIRCUIT WIRING

Branch circuit wiring (120 volts) runs from the main panel to the outlets and fixtures in the house. While inspecting the branch circuit wiring, the home inspector will identify the **type of conductor material** and the **type of cabling** carrying the wiring. The inspector compares branch wiring for compatibility of its **amperage rating** to fuses or breakers. The inspector is required to report any observed solid conductor **aluminum branch circuit wiring**.

The branch circuit wiring inspection includes inspecting the following within the house, garage, and exterior:

- Condition of the cable and the wires
- Proper size wiring for circuits
- Connections at the main panel
- Installation in unfinished areas
- Splices in the circuits
- Presence of handyman wiring

Types of Branch Wiring

In branch circuit wiring, the conductors or wires are typically copper, but the home inspector may find aluminum wiring in homes built in the 1960's to mid 1970's. Wires are wrapped in insulation. Today, wires are installed in homes in cables. A typical 15-amp household circuit today will have an insulated hot wire, the insulated neutral wire, and an uninsulated ground wire sheathed in a single cable. Special circuit wiring for some 240-volt appliances will have 4-wire cables — 2 hot wires, the neutral, and the ground wire.

The oldest type of wiring is **knob-and-tube wiring**, used until the 1930's or 1940's. Ceramic knobs were used to secure wire to structural surfaces; ceramic tubes were used to pass wires through wood framing members like floor joists and studs.

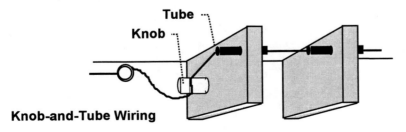

Knob-and-Tube Wiring

Guide Note

Pages 53 to 64 present information on the study and inspection of branch circuit wiring.

Definitions

Branch circuit wiring is that portion of the electrical system that runs from the electrical panel to outlets, switches, and fixtures in the home.

Knob-and-tube wiring is branch circuit wiring using ceramic knobs to secure wire to surfaces and tubes to pass wires through framing members.

Knob-and-tube wiring used rubber and cloth insulation around the wiring. Only 2 wires were used — a hot and a neutral — and they were strung separately, not encased together in a cable. No ground wire was used, so outlets used in knob-and-tube wiring have no grounding connection. When splices were made in the circuits, connections were made by twisting the wiring together, soldering the wires, and wrapping them in rubber, then in electrical tape. Today with modern wiring, connections must be made within closed junction boxes.

#17 Knob-and-tube wiring

Photo #17 shows an example of *knob-and-tube wiring* at the basement ceiling. Notice the ceramic tubes that pass through the floor joists and carry the wires through without touching the wood joists. Further back, you can see the knobs attached to the surface of the joists to secure the wires in position.

After knob-and-tube wiring went out of style, a metal armored cable, or MC or **BX cable**, began to be used. The wires were pulled through the cable during installation. When used in the 1930's to the 1950's, no ground wire was used. If connected to a 3-slot grounding outlet, the armored cable itself was depended on as the grounding conductor. Today, if BX or MC cable is used, the NEC requires that a ground wire be included.

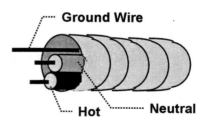

BX Cable
Metal Armored Cable

Worksheet Answers (page 52)

1. C
2. D
3. B
4. A
5. B
6. B
7. A
8. C
9. D
10. C

Almost all modern houses today use non-metallic sheathed cable or **Romex®** cabling. Romex® is a solid plastic jacket that carries 3 conductors for live, neutral, and grounding circuits. (Some of the very earliest versions of Romex were 2-conductor and didn't have the ground wire.)

Romex®
Non-Metallic Sheathed Cable

Electrical Metallic Tubing (EMT) may be used in places in household circuits. Special 240-volt circuits for large appliances such as stoves and

Electrical Metallic Tubing (EMT)

clothes dryers use a different third conductor, an additional hot wire. Often, the 3 conductors plus ground wires are carried to a kitchen to create a split outlet with 2 120-volt circuits at the outlet.

The home inspector should identify the types of cabling found in the home. There may be combinations of BX, Romex®, and conduit due to upgrading of the electrical service.

Wiring Practices

The National Electrical Code (NEC) has a list of rules governing the installation of branch circuit wiring. The following list covers many of the NEC requirements:

- **Bushings:** Where wires enter metal panels and boxes, there should be special bushings, grommets, or cable clamps to protect the wires from damage. The sharp edge of BX cable can cut through the wires' insulation unless protected with an approved bushing.

- **Surface wiring:** Where wiring exists on the outer surface of a wall or ceiling, it should be encased in conduit. Surface wiring may most often be seen in the basement and the garage or outside at the eaves.

- **Wood framing members:** Wiring should be installed so that wires cannot be mechanically damaged. When wiring runs perpendicular to the floor joists, it should be run

NEC WIRING RULES

- Proper bushings
- Enclosed surface wiring
- Protected by passing through wood framing members
- Stapled at boxes and secured over distance
- Covered junction boxes for connections
- Power to every room and area
- No extension-cord wiring
- Proper exterior cable
- GFCI protection

For Beginning Inspectors

Take the time to observe the wiring in exposed areas in your home. Start in the basement at the main panel and observe the cable leaving the box. See if you can tell which cables go to which areas of the house. Are some basement circuits in conduit? Can you tell which circuits are dedicated to large appliances and the furnace? Notice how the wiring is secured at junction boxes, switches, and outlets. Can you find any practices the NEC would not approve of?

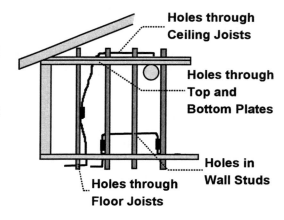

Holes through Ceiling Joists

Holes through Top and Bottom Plates

Holes in Wall Studs

Holes through Floor Joists

through holes drilled in the joists, not strung underneath the joists. Wiring parallel to a joist should be secured to the side of the joist and not attached on the underside of the joist. Ideally, wiring should pass through ceiling joists in the attic, although the home inspector may see cable secured to the top of the joists or rafters instead. Holes should be drilled through wall studs, top plates, and bottom plates and wires passed through to protect the wiring. Metal plates should be used at the outer edge of wall studs where protection is needed. This is generally done in those areas where nailing may occur, that is, higher up on the wall. The assumption is made that homeowners won't be hanging pictures a foot from the floor in the area of lower outlets.

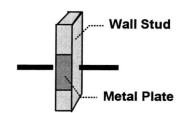

Wall Stud

Metal Plate

- **Secured wiring:** Wiring should be stapled to wood framing members or secured with hangers within 12" on both sides of all enclosures — panels, junction boxes, switches, outlets, and fixtures. This is true around ceiling fixtures as well. Wiring beyond the first secured point next to enclosures should be stapled or secured every 4 1/2' along the length of the circuit.

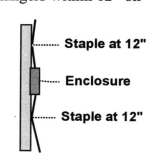

Staple at 12"

Enclosure

Staple at 12"

- **Covered junction boxes:** All connections made in the branch circuit wiring should be properly done and enclosed inside covered junction boxes. (Note that this was not the practice with knob-and-tube wiring. See page 54.) Any abandoned wires must be terminated properly in covered junction boxes as well.

- **Power available in each room and area:** It is required that there be a power source in every room or usable space in the home including stairwells, basements, crawl spaces, and attics.

- **No extension-cord wiring:** The NEC forbids the use of extension cords as permanent wiring. Extension cords should never be stapled to walls or floors. Their use is commonly seen in homes that don't have enough outlets providing power. It's common to see extension cords set up in basements or garages where not enough outlets are available.

- **Exterior cable:** Romex cable, type UF, is generally used underground. It should pass through conduit where it exits the ground and is attached to the structure. The NEC requires cable traveling horizontally to be buried, not left on the surface, and to be protected in concrete or conduit. Exterior connections require that splices be waterproof and enclosed in covered junction boxes.

- **GFCI protection:** Later NEC updates require the presence of GFCI's on all outlets within 6' of water as in the bathroom and kitchen, on exterior outlets and on garage outlets. More information will be provided about outlets, fixtures, and switches on pages 66 to 79.

Inspecting Knob-and-Tube Wiring

When the home inspector finds knob-and-tube wiring in a home, it should be recorded in the inspection report. The inspector should always **recommend that an electrician evaluate the wiring,** even if the inspector can't find obvious fault with the wiring. This wiring is just too old to accept as reliable without a professional opinion.

The NEC allows knob-and-tube wiring to remain in place while making proper extensions to the service, although many electricians will bypass it or remove it entirely. Generally, a home with this type of old wiring will not have enough outlets for modern living. Customers should be told that the wiring, even if in good condition now, will eventually have to be replaced.

During the inspection of knob-and-tube branch circuit

INSPECTING KNOB-AND-TUBE

- Damaged or brittle insulation
- Covered by thermal insulation
- Poor connections
- Amateur work
- Extension-cord wiring
- Use of "grounded" outlets
- 2-fuse circuits

"One of my inspectors was up in the attic of an old home with knob-and-tube wiring. The attic was insulated with old cellulose paper insulation. He actually saw sparks from the wiring falling into the insulation. He wanted to get out of there right away!

"The inspector left a very prominent note for the owners warning them of the extremely dangerous situation they had up in the attic and telling them that an electrician must be called in <u>today</u>. I'm not sure why anyone would ignore a warning like that, but a week later the home burned down.

"It's not unusual for the home inspector to be the one to discover safety hazards of this magnitude, but it is rare for people not pay attention to them."

Roy Newcomer

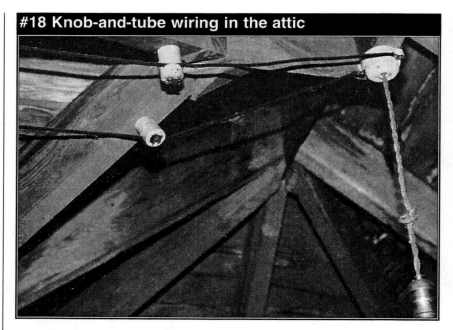

#18 Knob-and-tube wiring in the attic

wiring, the home inspector should look for the following conditions:

- **Damaged or brittle insulation:** The old rubber and cloth insulation can be damaged over time in exposed areas. In the attic, squirrels and mice can chew away the insulation and leave the wires bare. The insulation can become brittle from years of overheating or poor connections. Old copper wire can become brittle too. **These wires should *never* be buried under thermal insulation in the attic**.

- **Poor connections:** As was discussed on page 54, any original connections in knob-and-tube wiring were made by twisting, soldering, and wrapping the wires in rubber and electrical tape. Check any visible connections for their condition. And watch out for any recent extensions, upgrades, or repairs to the service. New connections should be properly made and encased in covered junction boxes. Some people seem to believe that since the original connections weren't done in junction boxes, the new ones don't have to be either. Wrong!

Connections can be especially poor at light fixtures. Most of them are the kind that you pull on to switch on. After years of pulling, the wires can come loose. Photo #18 shows an **old light fixture** in an attic. Notice the old cloth insulated wire at the fixture. We suggest to customers that these fixtures be replaced. These are often still found in the attic, basement, and in closets.

- **Amateur work:** A home with knob-and-tube wiring is hardly ever adequate for the needs of today's families. Somewhere along the line, someone will have extended the service in some way. Be on the look out for improper extensions to the service.

- **Use of "grounded" outlets:** The NEC forbids the use of 3-slot outlets when they aren't connected to the grounding circuit. And in knob-and-tube wiring, the switches, fixtures, and outlets are not grounded unless a separate ground wire was installed for each circuit during an upgrade to the service, which is not likely.

- **2-fuse circuits:** The home inspector may find the presence of 2 fuses on a single circuit, where both the hot wire and the neutral wire are fused at the main panel. If the neutral fuse blows, the circuit won't work, but the hot wire remains live with current. This is a **safety hazard**.

Inspecting Modern Wiring

When inspecting branch circuit wiring, the home inspector should pay attention to its source at the main panel and then observe all visible wiring in exposed areas in the basement, garage, attic, and exterior of the home. Watch for the following conditions:

- **Undersized wiring:** Always check that wiring is the appropriate gauge for the amperage of the fuse or breaker.

- **Improper installation, loose or drooping wiring:** As you trace wiring in exposed areas such as the basement, note whether NEC installation requirements are met. Note any wires stapled to the underside of joists, fixtures and boxes that are not securely attached on both sides, droops in long runs of cable, and the lack of conduit on surface wiring.

- **Damaged or worn wires and insulation:** Note where wires have been nicked, reducing their ability to carry current and causing overheating. Watch for any evidence of overheating and melted insulation. Report areas where rodents have nibbled on insulation, causing exposed wires.

BX cable can become corroded and worn. This should be reported. On the exterior, report any frayed or worn insulation.

- **Uncovered or missing junction boxes:** Report any unprotected splices in the circuit as a **safety hazard**. Splices should be properly executed and then enclosed in covered junction boxes. Uncovered boxes are also a safety hazard and should be reported as such. Be sure to inspect the junction boxes in the basement and garage areas for missing covers and exposed wires.

*Look at **Photo #19,** which shows an **uncovered junction box**. Here, the box is not only uncovered with wires exposed, but the wires aren't properly terminated. They're even hanging out. That's a serious safety hazard. Someone could reach up and touch those hot wires and get a nasty shock.*

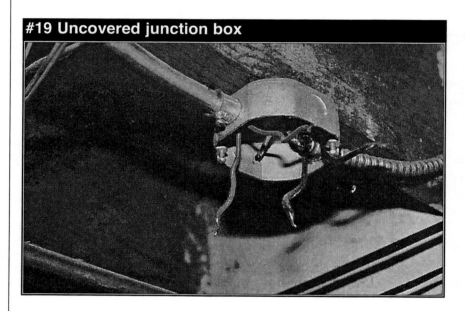
#19 Uncovered junction box

#20 Attic junction box

*Photo #20 is an **attic junction box**. Here's another serious safety hazard that must be reported. Not only is this box uncovered, but it's half buried in insulation, a terribly unsafe practice.*

- **Handyman wiring:** Be on the lookout for any amateur work in the branch circuit wiring. If the main panel looks like a spaghetti dinner, a warning alarm should sound for the home inspector. Illegal practices at the main panel can signal more amateur work on the circuits. There are so many inventive things handymen can try that it's hard to list them. Just be sure to watch for anything unusual. And always report any amateur wiring you find.

#21 Handyman wiring

*Photo #21 is one case of **handyman wiring** that we found. This photo is a reminder to you to look in closets during an inspection. Here, the homeowner has created another circuit by installing a porcelain light fixture on the closet floor with a fuse in it to supply more power to the room. That's inventive, but very dangerous. This situation is a fire waiting to happen. The wiring is sitting on the carpeting and is so close to the curtains. When you point out a safety hazard like this to your customers, you've earned your fee.*

- **Extension-cord wiring:** The home inspector often finds the use of extension cords as permanent wiring. This is most common in basement and garage work places, but can be seen in the house as well. Point this out to your customer and suggest that it be removed. If more circuits or outlets are needed, it's better to have an electrician in to do a professional job.

Personal Note

"One of my inspectors found a ceiling junction box with a loose cover in the basement. Thinking he'd be helpful, he started to screw the cover down tight. However, the box was not securely attached, and the box and wiring started to fall. He instinctively reached out to catch the falling box — with the screwdriver! The screwdriver connected with the current and delivered a good shock, throwing him about 5' backward.

"Don't do any electrical work during the inspection. Even the smallest and most helpful gestures can hurt you or do damage to equipment. It's much safer to point out the problem and leave it at that."

Roy Newcomer

Aluminum Wiring

Beginning in 1965 and continuing through the mid-1970's, aluminum was often used in branch circuit wiring as a replacement for scarce and expensive copper. However, there were serious fires reported as a result. Problems occur where small-gauge solid aluminum is used in 120-volt circuits (stranded aluminum in larger gauges is considered safe and still installed in new construction). The solid aluminum wires tend to expand and move out from terminal screws. Aluminum wire also tends to corrode at the connections. Both of these conditions create poor connections and serious overheating problems at outlets, switches, and at major appliances. (Copper-clad aluminum wire does not have these same problems.)

A house with aluminum branch circuit wiring can be made safer by fixing *every* connection to, or splice between, aluminum wiring in the home. The home inspector may find the following methods used to remedy the problem:

- **COPALUM crimp connectors:** The recommended method of repairing an aluminum-wired home is to bond copper wire to the aluminum wire at each outlet, switch, and junction. Here, a copper wire is attached to the existing aluminum branch wiring with a specially-designed metal sleeve and a powered crimping tool. The copper wire is then connected to the outlet or switch.

- **CO/ALR switches and outlets:** Another repair method is to replace all switches and outlets in the home with those labeled **CO/ALR**, which are designed to work better and safer with aluminum branch circuit wiring. This repair may be less than satisfactory since CO/ALR connectors are not available for all parts of the electrical system — for example, ceiling fixtures or permanently-wired appliances. Therefore, only part of the problem may be fixed.

- **Copper pigtailing:** Electricians will sometimes attach copper wires between the aluminum branch wiring and outlets and switches by a method called pigtailing. This is a connection made by twisting the wires together and covering them with a twist-on connector (wire nut). This method is not considered to be as good as the COPALUM crimp method, and it is felt that the pigtailing can fail just as easily as the original aluminum connection.

When the home inspector finds aluminum branch circuit wiring, an electrician should *always* be recommended to come in to evaluate the wiring. The inspector may find the presence of CO/ALR devices or see copper pigtailing repairs to such a service. But even then, an electrician should be recommended. It's just not possible for the home inspector to judge the safety of such an electrical service. And it's not a good idea to lead the customer to believe that the condition of aluminum branch circuit is safe when it might not be. The consequences are too severe.

Follow these steps when you find aluminum branch wiring in a home:

1. Always be sure to **report** the presence of aluminum branch circuit wiring in your inspection report.

2. **Explain** the history of aluminum wiring to the customer and explain why it may be a safety hazard. Inform the customer that repairs may have been made to the wiring, but that you can't determine if all connections have been repaired or if repairs have been made properly.

3. **Recommend** that an electrician be called to come to the house to evaluate the wiring completely and note this recommendation in the inspection report.

For Beginning Inspectors

If you have friends living in homes built from the mid-1960's to the mid-1970's, stop in to examine the electrical service. Start at the main panel and note whether aluminum branch circuit wiring was used. Trace some 120-volt circuits. Feel switches and outlets on these circuits for heat. In exposed areas see if you can find evidence of copper pigtailing, the use of COPALUM connections, and/or outlets/switches labeled CO/ALR.

Reporting Your Findings

Branch circuit wiring should be inspected where visible — on the exterior, in the garage, in the basement, and in the attic. Depending on the design of your inspection report, you may be reporting your findings on an electrical page or on separate pages for exteriors, garages, attics, and so on. In any case, you won't want to miss those items listed in the Don't Ever Miss list.

- **Branch wiring:** You should identify the type of branch circuit wiring found in the home (copper or aluminum) and the type of cabling found (knob-and-tube, BX, Romex, etc.). Report on as many types as you find. For example, a home may have old knob-and-tube wiring with upgrades using Romex cable. Note any damage, problems, and improper installation such as drooping wires in the basement.

- **Junction boxes:** Report any deficiencies in visible junction boxes if you find any, especially if they're uncovered or missing.

- **Recommending an electrician:** The conditions listed here are cause for recommending your customer have an electrician come in to evaluate the situation. Again, be sure to note this recommendation in your inspection report.

  — Knob-and-tube wiring
  — Aluminum branch circuit wiring
  — Wiring overheated, burned, damaged
  — Handyman wiring

- **Safety hazards:** The hazards listed here can be reported within the inspection report and then summarized on a summary page if your report has one. Make sure that customers understand what you're telling them about safety hazards. Show them where you've written about them in the report.

  — Undersized wires for circuit
  — Uncovered junction boxes with exposed wires
  — Wiring overheated, burned, damaged
  — Extension-cord or handyman wiring

WORKSHEET

Test yourself on the following questions.
Answers appear on page 66.

1. According to the standards, the home inspector is required to:

   A. Determine if knob-and-tube wiring is safe.
   B. Determine if metal protective plates are present on wall studs.
   C. Report any observed aluminum branch circuit wiring.
   D. Test connections at junction boxes.

2. Romex® cable is also called:

   A. Non-metallic sheathed cable
   B. Metal armored cable
   C. Rigid conduit
   D. Knob-and-tube

3. Identify the types of cabling shown here:

   A.

   B.

   C.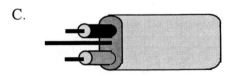

4. How should the home inspector report the condition found in Photo #19?

   A. The inspector should recommend that an electrician evaluate the service.
   B. The inspector should report the condition as a safety hazard.
   C. The inspector should report aluminum branch circuit wiring.

5. Where is the home inspector likely to find aluminum branch circuit wiring?

   A. In homes built before the 1930's
   B. In homes built during the 1940's and 1950's
   C. In homes built during the mid-1960's to mid-1970's
   D. In homes built during the 1980's

6. What condition causes aluminum branch circuit wiring to be unsafe?

   A. Overheating at poor connections
   B. Wire gauges too small for amperage
   C. Use of COPALUM connectors
   D. Absence of ground wires

7. For which of the following findings should an electrician be recommended to evaluate the service? *Circle as many letters as you wish.*

   A. Aluminum branch circuit wiring
   B. Drooping wiring
   C. Knob-and-tube wiring
   D. Extension-cord wiring
   E. Romex cabling
   F. Handyman wiring
   G. Wiring overheated, burned, damaged
   H. Uncovered junction box
   I. Use of conduit in surface wiring

8. Which of the following is a correct statement of an NEC requirement?

   A. Knob-and-tube wiring must be replaced and removed.
   B. The NEC forbids the use of extension cords even for temporary use.
   C. Exterior wiring should never be buried.
   D. Wiring should be secured to framing members within 12" on each side of boxes, switches, and outlets

Chapter Six

INSPECTING FIXTURES, SWITCHES, AND OUTLETS

Most standards of practice state the following requirements for the inspection of the switches, fixtures, and outlets inside the home and on the exterior.

- The inspector is required to **observe polarity and grounding** of all receptacles within 6' of interior plumbing fixtures, all receptacles in the garage or carport, and on the exterior of inspected structures.

- The inspector is required to **operate a representative number of installed lighting fixtures, switches, and receptacles** located in the house, garage, and exterior.

- The inspector is required to **operate all GFCI's**.

In general, this part of the inspection is to determine if power is available to each area of the property, if there are enough outlets in each area, if lighting is provided in each area and switches work, if outlets are properly wired and grounded, and if GFCI's are present and operable to protect people in those areas near water.

Lighting Fixtures and Switches

The NEC provides requirements for lighting fixtures and switches throughout the home, including the following:

- **Adequate amount:** The NEC requires at least one wall-switch-controlled lighting fixture in all rooms, hallways, stairways, utility rooms, basements, crawl spaces that are used for storage or that house equipment, plus attics, garages, and at outdoor entrances. **Stairways** with more than 6 steps should be switched both at the bottom and top and require 3-way switches.

- **Location:** Light switches should be located about 48" from the floor. Switches for fixtures in the **bathroom** should not be within reach of the tub or shower. Ceiling heat lamps in bathrooms should be beyond the swing of the door.

Worksheet Answers (page 65)
1. *C*
2. *A*
3. *A is BX cable.*
   B is conduit.
   C is Romex cable.
4. *B*
5. *C*
6. *A*
7. *A, B, C, D, F, G, H*
8. *D*

Lighting fixtures in **closets** should have a clearance all the way to the floor and should not be within 12" of the front edge of closet shelves. A recessed closet light should have a clearance of at least 6" from shelves. The latest NEC rulings banned the use of bare light bulbs in closets, requiring instead that fully enclosed fixtures be used.

- **Safety concerns:** Fixtures in showers and tubs should be waterproof. Exterior fixtures should be properly waterproof or weatherproof.

  When **recessed ceiling light fixtures** (IC or non-IC) are present on the top floor of the house, they should not be covered with insulation in the attic. A 3" clearance is required between recessed lights (non-IC) and attic insulation. IC lights can be covered with insulation.

These are the steps the home inspector should follow when inspecting lighting fixtures and switches:

1. Check for the **presence** of a wall switch and lighting fixture in each room, hallway, stairwell, attic, basement, crawl space, garage and outdoor entrances. Report the absence of lighting in any of these areas.

2. **Operate all lighting switches** and dimmers on the interior and the exterior to see if they work.

The home inspector should be on the lookout for the following deficiencies:

- **Old or missing fixtures:** The home inspector should recommend that old porcelain light fixtures used with knob-and-tube wiring be replaced. Take another look at **Photo #18**. This is an example of an old porcelain fixture. Tell your customers to have this replaced. The old cloth and rubber insulation on these fixtures is likely to be brittle and frayed.

  Note if any fixtures have been removed and the wires not properly terminated and covered.

- **Switches and fixtures that don't work:** When lights don't go on, it may be the result of a burned out bulb or blown fuse, but it could indicate a number of other conditions — defective wiring on the circuit or in the switch or fixture, a defective switch mechanism, or poor

---

INSPECTING FIXTURES AND SWITCHES
• Absence of lighting in any area
• Old and missing fixtures
• Switches and fixtures that don't work
• Warm, scorched, loose, damaged, or missing cover plates
• Unsafe practices
• Recessed lights covered by insulation

connections. The home inspector is required only to report that the switch and/or fixture are not working. But if lights flicker or buzz, then the cause of the problem is a loose connection which can become a shock or fire hazard. The customer should be advised that an electrician should check and repair the circuit.

- **Overheating at the cover plates:** When you flick on each switch, notice if there is any evidence such as heat, scorches, or signs of burning on the wall plate. This should be reported as a **safety hazard**, and the home inspector should recommend that an electrician be called to evaluate the problem.

  Report any **loose, damaged, or missing** cover plates which might allow a person to come in contact with live electrical wires.

- **Unsafe practices:** Report light switches too close to tubs and showers, closet lights too close to shelving, exterior fixtures that aren't waterproof, and any other unsafe practices you may find.

*Photo #22 shows a **ceiling fixture**. Take a good look at it. Here, the homeowner has simply pulled out the wiring, drooped it over a hook in the ceiling, and connected it to a receptacle for a light bulb. There is no box or plate protecting these connections. This situation was reported as a **safety hazard**.*

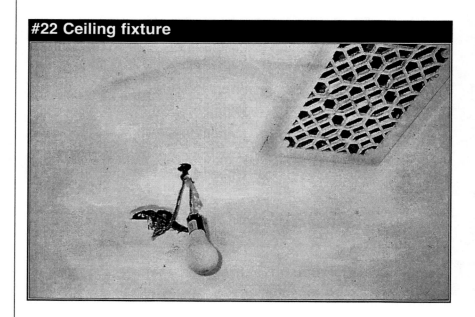

#22 Ceiling fixture

- **Recessed lights covered by insulation:** Whenever you find recessed ceiling lights in the top floor of the house, make a note to check it out when you get into the attic to see if there is the proper clearance around that light. If you don't remind yourself, you may miss those lights that are entirely covered with insulation. Tell the customer that either the homeowner or the customer after purchasing the house should remove the insulation from around the light to check if the light should have clearance or if it is rated for an insulated ceiling. The home inspector does not have to dig out around recessed lights. Simply report that the lights are covered with insulation and it is a **safety hazard** unless properly rated.

INSPECTION ORDER

- Exterior service entrance, fixtures, and outlets
- Garage wiring and outlets
- Interior power sources, fixtures, and outlets
- Attic branch circuit wiring
- Basement main panels, subpanels, branch circuit wiring, outlets

#23 Buried recessed light

Photo #23 shows a *buried recessed light*. In this case, it appears that the light was at one time baffled to keep insulation away, but it's now covered. We had no obligation to dig out the insulation to check the rating on the light. We did, however, inform the customer that the situation was a possible safety hazard and could start a fire. We suggested that someone check the rating and/or remove the insulation from the around the light.

NOTE: The home inspector may find unusually shaped rectangular or oval switches for the lighting system. This may be an indication of the presence of a **low voltage lighting system.** As discussed earlier, this type of lighting system was popular in the 1950's. It used 12- or 24-volt wiring, instead of 120-volt, boxes of relays, and one or more switching panels that controlled all the lights in the house. The home inspector should inspect the presence of light sources in each area and see if the switches work, but does not have to otherwise inspect such a system.

Definitions

A <u>grounded outlet</u> is one which is wired with a ground wire that is connected to the grounding system.

A <u>split outlet</u> is wired with 2 hot wires and 1 neutral to provide 2 separate circuits to the outlet plus the ground wire for grounding purposes.

A <u>GFCI outlet</u> is one which has a monitoring device installed that will trip the circuit when a ground fault is detected. GFCI stands for ground fault circuit interrupter.

Electrical Outlets

Electrical outlets provide receptacles where appliances can be plugged into a circuit. Most have brass screws on one side to which the live wire (black) is connected. The other side has silver-colored screws to which the neutral (white) wire is connected. These 2 wires provide a 120-volt circuit to the receptacles in the outlet. If the outlet is grounded, the ground wire is connected to a screw toward the bottom of the outlet.

Until 1960, most electrical outlets were **ungrounded**. They had only 2 slots in them for appliance plugs with 2 prongs. Some of them were **polarized** — that is, they had a larger neutral slot and a smaller hot slot so polarized appliances with large and small

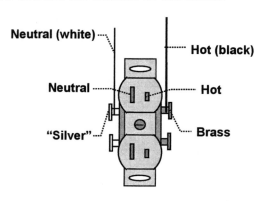

UNGROUNDED OUTLET

prongs could be plugged in correctly. The hot wire, brass screw, and small slot is an industry convention. (The older types of ungrounded outlets had slots the same size.)

Some 2-slot outlets are not grounded. When these old outlets are replaced, 3-slot outlets are typically used. That poses a problem because they may still be lacking a ground wire to the outlet. The home inspector cannot tell if a replacement 3-prong outlet is grounded without testing it.

NOTE: In 240-volt circuits, a second hot wire (typically red) would be wired to the outlet in addition to the neutral wire.

After 1960, most outlets were **grounded**. A ground wire provides a means of escape for electricity if something goes wrong with the outlet or the appliance. Any live current is then channeled to the grounding system rather than into the person touching the outlet. Each receptacle has 3 slots — the large neutral, the small hot, and a little round oval below the others. The 3-slot

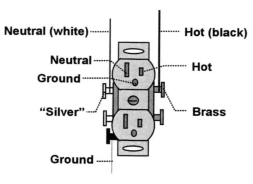

GROUNDED OUTLET

grounded outlets are of value only where appliances with 3-prong plugs are used. Appliances such as refrigerators, washers and dryers, microwaves, computers, and power tools are examples. A 2-prong plug does not connect to the grounding slot.

Another kind of outlet used today is the **split outlet**. This type of outlet has 2 complete 120-volt circuits in a single outlet.

Such outlets are common in kitchens where more power is needed. In fact, most kitchen counter outlets today are split. An appliance such as a coffee maker plugged into the top receptacle is on a different circuit from a toaster plugged into the bottom one.

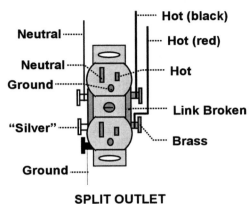

SPLIT OUTLET

In a split outlet, a second hot wire (red) is connected to the lower brass screw. The link or connection between the top receptacle and the bottom is broken or removed to separate the circuits.

Another type of outlet is the **GFCI outlet**. The GFCI, or ground fault circuit interrupter, is a monitoring device that compares the amperage of the current flowing to and from the outlet. A leak at the appliance or outlet will be detected. The leak may be too small to blow a fuse or trip a breaker. The GFCI will trip in 1/40 second after a ground fault of only .005 amps is detected, interrupting the flow of electricity to the circuit. A GFCI on the first outlet of a circuit will protect all downstream outlets on the same circuit. When the GFCI trips, power is interrupted to all downstream outlets as well.

GFCI's are often added to ungrounded circuits to provide protection, and some local codes allow this as an alternative to grounding.

GFCI's should be tested monthly to be sure they're still in working condition. When the **test button** on the GFCI is pushed, the **reset button** should pop out, indicating that the GFCI is working. At this point, the power is interrupted to the outlet. Power is reinstated by pushing the reset button back in.

OUTLET LOCATIONS

- Every room, basement, and garage

- Every 12' along interior walls and 12" above floor

- Present in hallways longer than 10'

- Every 4' along kitchen countertops

- At least 3' from tub or shower

NOTE: Appliances that could be damaged or cause unreasonable inconvenience if power is lost should not be plugged into GFCI outlets. And GFCI's don't do well on circuits for motor-driven appliances because they can become damaged by motor surges. Computers, security systems, and appliances like refrigerators and freezers shouldn't be plugged into GFCI outlets.

Outlet Requirements

The NEC requirements for electrical outlets throughout the home include the following:

- **Grounding:** Today, all new construction requires grounded outlets. For older homes, the NEC prohibits the use of 3-slot outlets when there is no grounding system.

  Location: Outlets are required every 12' along interior walls and in hallways that are 10' or longer. Wall outlets are generally set at 12" from the floor, but they should not be present over electrical baseboard heating units. Kitchen countertop outlets should be vertical (not horizontal) and be spaced no more than 4' apart. There must be a bathroom outlet near the sink but no more than 3' from each basin. Outlets are required in every room, including utility rooms, and in the basement and garage. The NEC requires outlets to be flush with the surface of a flammable wall surface or recessed up to 1/4" if the wall surface is non-flammable.

- **Dedicated circuits:** Dedicated circuits are required for large appliances such as the dishwasher, refrigerator, freezer, and disposal. The NEC requires at least 2 20-amp small appliance circuits for the kitchen. These outlets must be rated for the amperage of the circuits. GFCI's are not required for dedicated circuits.

- **GFCI's required:** The NEC requires GFCI's for all **exterior outlets**, all outlets **within 6' of water** such as those in the bathroom, wetbar, or kitchen (except for those dedicated to large appliances) and accessible **garage outlets**. Additionally, all outlets that serve counters in the kitchen must be GFCI protected.

- **Exterior outlets:** All exterior outlets are required to be weatherproof.

Inspecting the Outlets

These are the steps the home inspector should follow when inspecting electrical outlets:

1. Check for the **presence** of an outlet in each room, long hallways, basement, and the garage. Report the absence of a power source in any of these areas.

2. Check for the **presence of GFCI's** in areas that require them and **operate all GFCI's** test buttons <u>and</u> test them with the GFCI tester to see if they work.

3. Check at least one outlet per room for **current**, and those within 6' of water, in the garage and outside for **current, polarity**, **open ground**, and **correct wiring** with the GFCI tester and/or neon bulb tester.

The home inspector should inspect outlets for the following:

- **Insufficient outlets:** Check for the presence of an outlet in each room and be sure to report if any room doesn't have the required outlet. Many a home inspector has had to pay to wire a room where he or she missed reporting that the room didn't have a power source. Note areas such as basement or garage workshops with extension-cord wiring, noting the lack of enough outlets in the area. Check in the kitchen too for extender outlets and extension cords that indicate that more outlets are needed. Point out to the customer that the homeowner appears to need more outlets and that the customer should expect the same problem.

- **Overheating at the cover plate:** Feel and look at the outlet covers for any evidence of overheating such as heat, scorches, or signs of burning. This should be reported as a **safety hazard**, and the home inspector should recommend that an electrician be called to evaluate the situation.

    Report any loose, damaged, or missing outlet covers which might allow a person to come in contact with live wires.

- **Unsafe practices:** Report outlets too close to tubs and showers, extension-cord wiring, outlets too close to heat sources, and outlets not placed according to NEC rules.

INSPECTING OUTLETS

- Insufficient outlets
- Warm, scorched, loose, damaged, or missing cover plates
- Unsafe practices
- Ungrounded outlets
- GFCI's missing or not operating
- Incorrect wiring

---

*Photo #24 shows an **unsafe situation**. Here, the outlet has been placed over an electric baseboard heat register. Having outlets over heat sources can be dangerous because a cord draped over it can start on fire. Anytime you find an outlet too close to heat, make a note of it in your inspection report and inform customers that wires plugged into these outlets can be a **safety hazard**.*

- **Ungrounded outlets:** Some homes have ungrounded outlets and that's not a problem in itself. However, the home inspector might find some unsafe things going on at these outlets that can be pointed out to the customer.

  These 2-slot outlets will sometimes have an **adapter** in them which has a 3-slot receptacle for the appliance plug and a 2-prong plug to go into the outlet. A grounding connection on the adapter is connected under the plate screw, supposedly to provide grounding. However, if the outlet is not grounded, the adapter won't be grounded either. Sometimes, people break off the third prong of a 3-prong plug and plug it into a 2-slot ungrounded outlet. This shouldn't be done and is unsafe.

  Ungrounded 3-slot outlets should have the ground slot filled with an epoxy or a pin designed for this use so no 3-prong plug can be used in it or replaced with a 2-slot outlet.

- **GFCI's missing or not operating:** Check that all required outlets (within 6' of water in bathroom and garage wall outlets, exterior outlets, and kitchen as required) have GFCI's and report them if they do not. Next, before you test each one for operation, it's a good idea to ask the owner what other rooms may be on the same circuit. It's generally okay, because computers and security systems aren't usually on the same circuits as bathrooms and kitchens. But you might trip the upstairs bathroom outlet if bathrooms are on the same circuit. *Be sure to reset all GFCI's.*

Now, test each GFCI for operation *twice* with the following methods:

1. Insert the **GFCI tester** into the outlet. If the circuit is correctly wired, the correct lights on the tester will be on. Then **push in the black button** on the tester. This should trip the GFCI and interrupt the circuit, and all lights on the tester will be off. If the 2 lights remain on, either the GFCI is not working or the circuit is incorrectly wired. If the GFCI worked, push in the reset button on the GFCI outlet. (Note: Not all GFCI testers have the same light orientation. Be sure to read directions for your tester.

Lights Off
GFCI Functioning

Lights On
Not Functioning

2. Next, push the **test button on the outlet** to test it again. If the reset button pops out and the tester lights go off the GFCI is working and wired properly. Push the reset button on the GFCI outlet to reset the circuit.

CAUTION: You may find large appliances such as refrigerators, freezers, or washers plugged into GFCI circuits when they shouldn't be. Freezers and refrigerators in the garage plugged into GFCI outlets are a special problem. After the home inspector trips that GFCI, the freezer will have to restart and the motor surge can trip the GFCI again after the inspection. That means spoiled food and a complaint to the home inspector. If you come across this situation, leave a note for the owners informing them to check that the GFCI hasn't tripped after the inspection.

- **Incorrect wiring:** See the discussion that follows on this page and the next.

Personal Note

"I've tested GFCI's in several garages that had freezers plugged into them. Very often the GFCI will trip again after the inspection and I'll get a call about spoiled food. And let me tell you, the freezers involved in my inspections always seem to be holding some very exotic and expensive food like salmon from Norway. Paying to replace food like that can get expensive.

"Now I always leave the owner a note cautioning them to watch the freezer because the GFCI may trip again. I also tell the owner the freezer shouldn't be plugged into a GFCI outlet. It saves me money."

Roy Newcomer

Reverse polarity is where the hot wire is wired to large slot in an electrical outlet and the neutral wire is wired to the small slot, the opposite of how it should be done.

An open ground means that the outlet is not properly grounded. A 2-slot outlet may not be grounded, while the 3-slot outlet should be grounded.

Testing the Outlets

We've already talked about testing GFCI outlets for operation on page 71. The home inspector should also test:

- One outlet in each room to see if it is operative or not

- Interior outlets within 6' of water and those in the garage and outside for correct wiring, reverse polarity, open ground, and whether they are inoperative

Any **2-slot outlets** will have to be tested with the **neon bulb tester**. This is the small tool with a light attached between 2 long prongs that can be inserted in the outlet slots and against the outlet screw to get readings. The table below shows how to test 2-slot outlets with the neon bulb tester.

2-Slot Outlets	
Neon Bulb Tester Off = ⬬ On = ⬭	**Condition**
	Correct wiring where hot wire is wired to small slot and neutral wire is wired to large slot.
	Reverse polarity where hot wire is wired to large slot and neutral wire is wired to small slot.
	Inoperative for any number of reasons but basically not providing power to the outlet.
	No ground where checking small and large slots shows that outlet has power, but checking each slot against screw shows _no_ ground.

Test **3-slot outlets** with both the **GFCI tester and the neon bulb tester**. This is a double check of polarity, open ground, and inoperative outlets.

3-Slot Outlets		
GFCI Tester Off = ● On = ○	**Neon Bulb Tester** Off = ◖ On = ◗	**Condition**
		Correct wiring where hot wire is wired to small slot and neutral is wired to large slot.
		Reverse polarity where hot wire is wired to large slot and neutral to small slot.
		Open ground where the outlet is not grounded and no current flows between the upper slots and the grounding slot.
		Inoperative for any number of reasons, but basically not providing current or power to the outlet.
		Many conditions but usually means there is a **weak ground** which indicates a leak of current _through_ the outlet.

Be sure to test all **outlets near water** for polarity and open grounding. Outlets near water with these problems can be especially dangerous and should be reported as a **safety hazard** if found. Check the outlet near the main panel for grounding; sometimes this outlet isn't grounded.

Don't ever miss noting the absence of power to any area or checking at least one outlet per room, including the utility room, basement area, the garage, and all the exterior outlets.

TESTING OUTLETS

- Correct wiring
- Reverse polarity
- Open ground
- Inoperative

For Beginning Inspectors

This is a good time to buy yourself a GFCI tester and neon bulb tester and get some practice. (Note: Not all GFCI testers have the same light orientation. Be sure to read directions for your tester.)

GFCI Tester

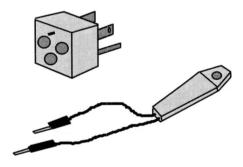

Neon Bulb Tester

Use these tools to test all your own outlets first. Then carry the testers with you for several days and test outlets at other people's homes. Always be aware that you may be turning off power to important outlets downstream of GFCI's. Ask first.

Putting It All Together

This Practical Guide has presented the inspection of the electrical system by components, not by the flow of how the electrical inspection would be performed. Let's review how your inspection would really happen.

1. **Exterior electrical:** While inspecting the exterior of the house, you would inspect the service entrance, including your check of the mast, masthead, cabling, and meter or main panel. You would take note of clearances of the service drop and note the system voltage rating by the number of service conductors entering the masthead. As you circle the house, you test all exterior outlets for polarity and open grounds, note the presence and test all GFCI's, and note weatherproofing and proper enclosures for fixtures.

2. **Garage:** Be on the lookout for handyman or extension-cord wiring. Check that surface wiring is enclosed in conduit. Note the presence and operation of GFCI's. Check to see if there's a freezer or refrigerator plugged into a GFCI outlet.

3. **Interior electrical:** In the kitchen and bathroom and in any area where outlets are near water, check for the presence and operation of GFCI's, polarity, and open ground. In all other rooms and areas, check at least one outlet per room for power. Note the absence of power to any room. Note proper placement and condition of outlets and fixtures. Touch cover plates to check for overheating.

4. **Attic area:** Here's where you'll get your first look at branch circuit wiring. Remember to report the existence of knob-and-tube wiring. Watch for wiring and recessed lights buried in insulation and note open junction boxes. Watch for evidence of overheating at connections.

5. **Basement:** Here, you'll probably find the main panel and subpanels servicing the home. Remove covers, if safe, and examine them for determination of system amperage, compatibility of components, grounding, and type of branch circuit wiring. Remember to report the presence of aluminum branch circuit wiring. Watch for deficiencies at the panels — handyman wiring, tapping before the main, double tapping fuses or breakers, arcing or burned back wiring, melted insulation, and evidence of undersized fusing or wiring.

Trace branch circuit wiring in the basement area and note dedicated circuits to large appliances, junctions boxes, fixtures, and presence of outlets. Watch for proper use of conduit and cabling, extension-cord wiring, open junction boxes, and so on.

Reporting Your Findings

Now, let's get back to reporting your findings from the inspection of **fixtures, switches, and outlets**. Your inspection report should have the appropriate places in which to report these findings. Reports that organize these electrical elements of the inspection with the location (exterior, garage, kitchen, bathroom, other rooms, attic, and basement) are easiest to fill in.

In general, you want to be able to record the presence or absence of lighting and power in each area. You should report the presence or absence of GFCI's and their operation in areas where they should be found. Also, you should report the results of your tests for reverse polarity and open grounds.

Record your comments on other important findings such as these examples:

- "Missing cover on junction box on south exterior wall."
- "Recommend GFCI's for garage and bathroom outlets."
- "Should add outlets in attic and remove extension cords."
- "Have electrician fix polarity on outlet near kitchen sink."
- "Hot switch plate, should have electrician examine."
- "Closet light fixture too close to shelving."

Be sure to record the following **safety hazards** if you've found them. Remember, it's a good idea to repeat these on a summary page of your inspection report:

- Reverse polarity or open grounds near water
- Evidence of overheating at fixtures, switches, and outlets
- Unsafe practices such as outlets too near tub or shower

DON'T EVER MISS

- Absence of lighting or power in any room or area
- Switches, fixtures, and outlets that don't work
- Warm, scorched, loose, damaged, or missing cover plates
- Unsafe practices
- Recessed lights covered by insulation
- Reverse polarity or open grounds near water
- GFCI's missing or not operating

WORKSHEET

Test yourself on the following questions.
Answers appear on page 88.

1. Which of the following is <u>not</u> a NEC requirement?

   A. One lighting fixture in all rooms, hallways, stairways, basements, attics, garages, and outdoor entrances
   B. Outlets in all rooms, basement, and garage
   C. All new construction with grounded outlets
   D. Dedicated circuit for each outlet

2. Outlets at sinks in the bathroom should:

   A. Have GFCI's.
   B. Be located at least 5' away from tub and shower.
   C. Be located at least 3' away from tub or shower.
   D. Be waterproofed.

3. Electrical outlets should be present:

   A. Above electrical heat registers.
   B. At least 48" from the floor
   C. Every 12' along interior walls
   D. Every 12' along kitchen countertops

4. According to the standards, the home inspector is required to:

   A. Operate all GFCI's.
   B. Operate and test all lighting fixtures, switches, and outlets.
   C. Test for reverse polarity and open grounds on the exterior outlets only.
   D. Rewire outlets with reverse polarity.

5. Which of the following should be reported as a safety hazard?

   A. Insufficient outlets
   B. No lighting in utility room
   C. Inoperative outlet
   D. Recessed light buried in insulation

6. When testing 2-slot outlets with the neon bulb tester, which of the following would show reverse polarity?

   A. B.

7. When testing 3-slot outlets with the GFCI tester, which of the following would show an open ground?

   A. B.

   C. D.

8. What is a <u>likely</u> condition that could be present with the following results of testing a 3-slot outlet with the neon bulb tester?

   A. Correct wiring
   B. Reverse polarity
   C. Weak ground
   D. Open ground

   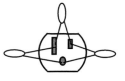

9. What is a split outlet?

   A. One that has a complete circuit for each receptacle
   B. One that is only partially grounded
   C. One that has a dedicated circuit
   D. One that is located outside

10. Freezers should be plugged into a GFCI outlet.

    A. True
    B. False

EXAM

A Practical Guide to Inspecting Electrical has covered a great many details involved in the inspection of the exterior of a home. Now's the time to test yourself and see how well you've learned it. I included this exam in the guide so you'll have that chance, and I hope you'll try it.

To receive Continuing Education Units:
Complete the following exam by filling in the answer sheet found at the end of the exam. Return the answer sheet along with a $50.00 check or credit card information to:

American Home Inspectors Training Institute
N19 W24075 Riverwood Dr., Suite 200
Waukesha, WI   53188

Please indicate on the answer sheet which organization you are seeking CEUs.

It will be necessary to pass the exam with at least a 75% passing grade in order to receive CEUs.

Roy Newcomer

Name_____

Address_____

_____

Phone:_____

e-mail:_____

Credit Card #:_____

Exp Date:_____

Fill in the corresponding box on the answer sheet for each of the following questions.

1. Which action is required by most standards of practice?

   A. Required to observe polarity and grounding of all interior outlets within 6' of water
   B. Required to use a GFCI tester and neon bulb tester at the main panel
   C. Required to determine if aluminum branch circuit wiring is safe
   D. Required to dismantle the main disconnect and main overload protection devices

2. Which of the following, according to most standards, must be inspected in an electrical inspection?

   A. CO detectors
   B. Low voltage systems
   C. Branch circuit wiring
   D. Security system wiring

3. What is the overall purpose of the electrical inspection?

   A. To determine the electrical service's amperage rating
   B. To identify the type of service conductors
   C. To identify major deficiencies in the electrical system
   D. To identify safety hazards in the service

4. How do the standards describe the inspection of a home's electrical system?

   A. As an exhaustive technical inspection
   B. As an NEC code inspection
   C. As a local code inspection
   D. As a visual inspection

5. What steps should the home inspector take to protect the customer's safety?

   A. Ask the customer not to be present during the electrical inspection.
   B. Have the customer stay a safe distance away from electrical equipment.
   C. Turn off all power to the house so the customer can't be harmed.
   D. Don't inspect any portion of the service that could produce a shock.

6. What is resistance?

   A. Flow of electricity
   B. Opposition of a material to the flow of electricity
   C. Force that drives electrons in a current
   D. Heat produced by the flow of electricity through a material

7. What is the measure for electromotive force?

   A. Volts
   B. Amps
   C. Ohms
   D. Watts

8. A material with a high resistance would be a:

   A. Good conductor
   B. Good transformer
   C. Good neutral wire
   D. Good insulator

9. Which formula shows the relationship voltage, amperage, and wattage?

   A. A = V x W
   B. W = V x A

10. What amperage would be needed to power a 1200 watt hair dryer on a 120 volt system?

    A. 1 amp
    B. 100 amps
    C. 10 amps
    D 0.1 amps

11. What are the NEC-required service drop clearances for A, B, C, D indicated in this drawing?

    A. 3 ft., 10ft., 12ft., 3 ft.
    B. 18 ft., 12 ft., 10 ft., 3 ft.
    C. 3 ft., 8 ft., 12 ft., 10 ft.
    D. 18 in., 10 ft., 12 ft., 3 ft.

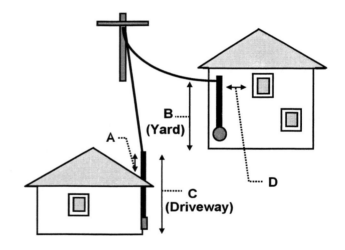

12. What would be the voltage rating of an electrical service with the condition at the masthead as indicated in the drawing below?

    A. 120-volt service
    B. 120/240-volt service

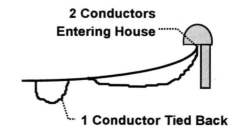

2 Conductors Entering House

1 Conductor Tied Back

13. Which electrical meter most likely indicates the largest service amperage?

 A. One with a round base
 B. One with a square base
 C. One with a rectangular base
 D. One without a base

14. Which 2 types of service conductor require the same gauge for a given amperage rating?

 A. Copper and copper-clad aluminum
 B. Copper and aluminum
 C. Aluminum and copper-clad aluminum

15. Which of the following photos shows a pull-out fuse block main disconnect?

 A. Photo #3
 B. Photo #6
 C. Photo #7
 D. Photo #9

16. What is the amperage rating of the main disconnect in this drawing?

 A. 60 amps
 B. 100 amps
 C. 120 amps
 D. 200 amps

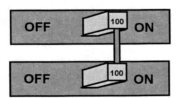

17. What would be considered a safety hazard at the main disconnect?

 A. The use of fuses instead of breakers
 B. Aluminum service conductors
 C. 3 hand movements required to turn off
 D. 60-amp conductors and 100-amp main breakers

18. Tapping before the main disconnect is considered an unsafe practice.

 A. True
 B. False

19. **Case study:** You find the incoming service conductors to be #6-gauge copper. The main pull-out fuse block contains 2 100-amp fuses. The main panel is rated for 60 amps, and the meter base is round. What do you report as the service amperage rating?

 A. 30 amps
 B. 60 amps
 C. 100 amps
 D. 200 amps

20. In the case study above, what should be reported as a safety hazard?

 A. The service conductors are rated at a higher rating than the main fuses
 B. The service conductors are rated at a lower rating than the main fuses

21. Identify the components marked 1, 2, 3 in the drawing of a single-bus main panel:

 A. Sub-panel breaker, bussbar, grounding bar
 B. Grounding bar, bussbar, sub-panel breaker
 C. Main breaker, bussbar, grounding bar

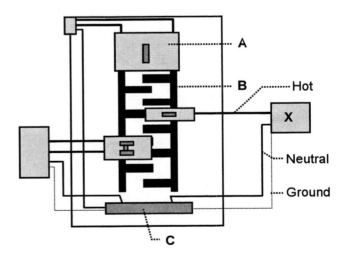

22. In the drawing above, what size circuit would a load at the letter **X** be on?

 A. 120-volt circuit
 B. 240-volt circuit

23. Which method of subpanel wiring is not acceptable?

   A. Feeder wires from special lugs in the main panel
   B. Feeder wires tapped at the main disconnect in the main panel
   C. Feeder wires from a 240-volt breaker in the main panel

24. Which condition would not stop a home inspector from inspecting a main panel?

   A. If water is pooled on the floor below the panel
   B. If arcing is heard from within the panel
   C. If heat is felt when touching the cover
   D. If headroom at the panel is only 6'

25. What should the inspector suggest to the customer if the main panel is underrated for the service amperage?

   A. Replace the panel.
   B. Have an electrician evaluate the service.
   C. Use fuses and breakers to match the panel rating.
   D. Add a subpanel to get more circuits.

26. When inspecting a main breaker, the home inspector should:

   A. Turn it off to see if it turns off all power to the house.
   B. Jiggle the service conductors to see if they're tight.
   C. Compare its amperage rating to service amperage.
   D. Remove it from the panel to inspect it.

27. What is an unused knockout?

   A. 2 breakers without a connecting handle
   B. A GFCI that doesn't trip the circuit
   C. A breaker panel with a space available without a breaker in it
   D. Multiple tapping on the main lugs

28. How does a home inspector test a GFCI at the main panel?

   A. Flip the breaker off and see if the test button pops, then flip the breaker back on.
   B. Push the test button and see if the breaker flips off, then flip the breaker back on.
   C. Flip the breaker off and then back on.
   D. Use the GFCI tester.

29. For which of the following conditions should an electrician be called to evaluate the service?

   A. Handyman wiring at the main panel
   B. A rusted main panel
   C. The presence of a 20-amp fuste on a 15-amp circuit
   D. All of the above

30 Which of the following conditions at the main panel should be reported as safety hazards?

   A. Inoperable GFCI breaker, double tap of main wires, oversized fuses/breakers, uncovered panel with exposed wires
   B. Unlinked breakers, presence of split buss, inoperable GFCI breaker, oversized fuses
   C. Double tap of main breaker, inoperable GFCI breaker, 4 hand movements to turn off power, oversized fuses
   D. Oversized breakers, double tap of main breaker, presence of a split buss, system not grounded

31. Which photo is an example of handyman wiring?

   A. Photo #2
   B. Photo #8
   C. Photo #13
   D. Photo #16

32. Which of the following actions is not a requirement regarding branch circuit wiring?

    A. Required to report aluminum branch circuit wiring

    B. Required to observe branch circuit conductors

    C. Required to observe compatibility of branch circuit wiring and their fuses or breakers

    D. Required to dismantle connections at junctions boxes

33. Which of the following is used as branch circuit wiring in most modern homes?

    A. Romex cable

    B. Conduit

    C. BX cable

    D. Knob-and-tube wiring

34. Which of the following is NOT an NEC requirement?

    A. Connections in branch circuit wiring should be enclosed in junction boxes

    B. All outlets should have GFCI protection

    C. Wiring should be run through wood framing members rather than stapled to the edges.

    D. There should be a power source in each room

35. Why should knob-and-tube wiring never be buried under thermal insulation?

    A. It's not grounded.

    B. Wires can become bare from rodents nibbling on them and from worn insulation.

    C. Knobs can come loose if someone steps on them.

    D. There can be loose connections at light fixtures.

36. What is an example of improper installation of branch circuit wiring?

    A. Connections not in junction boxes

    B. Wires stapled to the sides of joists

    C. Surface wiring in conduit

    D. Using 3-slot outlets that are grounded

37. BX cable is also called:

    A. Rigid conduit

    B. Non-metallic sheathed cable

    C. Metal armored cable

    D. Knob-and-tube wiring

38. What should the home inspector do if he or she finds aluminum branch circuit wiring?

    A. Recommend it be replaced

    B. Not inspect it

    C. Recommend an electrician be called in to evaluate the electrical system

    D. Suggest that the customer shouldn't purchase the home

39. What is the most highly recommended method of repairing aluminum branch circuit wiring?

    A. Pigtailing copper to the aluminum wiring with wire nuts at every connection and splice in the wiring

    B. Use of CO/ALR devices

    C. Using COPALUM crimp connectors to bond copper to the aluminum wiring at every connection and splice

40. Which of the following conditions of circuit wiring should be listed as a safety hazard?

    A. Copper circuit wiring, undersized wires for circuit, wires overheated/burned

    B. Undersized wires for circuit, uncovered junction boxes with exposed wiring

    C. Handyman wiring, copper circuit wiring, undersized wire for circuit

    D. Undersized wires for circuit, uncovered junction boxes with exposed wired, aluminum main wires

41. The home inspector is NOT required to:

   A. Test all outlets in the house for open grounds and polarity

   B. Operate all GFCI's

   C. Check for lighting and power source in each room

   D. Check all rooms for AFCI protection

42. Which of the following is an unsafe practice?

   A. A bathroom outlet at the sink with a GFCI
   B. A shower with a recessed lighting fixture
   C. A bathroom wall switch 6' from the tub
   D. A bathroom outlet 12" above electric heating unit

43. In outlet wiring, to which screw is the hot wire connected?

   A. Silver-colored screw
   B. Brass screw
   C. Ground screw

44. In outlet wiring, which slot is hot?

   A. The small slot
   B. The large slot
   C. The little oval slot

45. Which of the outlets shown here is wired as a split outlet?

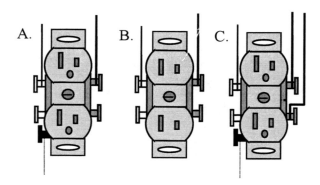

46. Every outlet on a property should be a GFCI outlet.

   A. True
   B. False

47. **Case study:** You are inspecting a home's electrical service. The service conductors are aluminum; the house wiring is copper. Photo #15 shows the condition of the main panel. Photo #19 shows what you found in the basement. Fixtures, switches, and outlets are mostly in satisfactory condition. However, there are no GFCI outlets in the house, the kitchen has only one countertop outlet, and the outlet near the kitchen sink tests for reverse polarity.

For what condition, if any, would you recommend that an electrician evaluate the electrical service?

   A. Aluminum branch circuit wiring
   B. Handyman wiring
   C. Would not recommend an electrician

48. In the case study above, what conditions would you report as a safety hazard?

   A. Handyman wiring, open junction box with wires exposed, reverse polarity in kitchen

   B. Handyman wiring, insufficient outlets in kitchen, aluminum service conductors

   C. Handyman wiring, open junction box with exposed wiring, insufficient outlets in kitchen

   D. Open junction box with wires exposed, reverse polarity in kitchen, aluminum service conductors

49. In the case study above, what recommendations would you make to the customer?

   A. Remove open junction box in basement, have electrician evaluate the service, add more outlets in the kitchen

   B. Have electrician evaluate the service, replace service conductors with copper, have kitchen outlets rewired

   C. Have electrician evaluate the service, install GFCIs on interior outlets near water, add more outlets to kitchen

   D. Install GFCIs on interior outlets near water, remove open junction box in basement, have kitchen outlet rewired.

50. Aluminum branch wiring was
recommended as a substitute for scarce and
expensive copper wire approximately when:

A. 1940 – 1950

B. 1965 – 1975

C. 1975 – 1985

D. 1955 – 1965

Worksheet Answers (page 80)

1. D
2. A
3. C
4. A
5. D
6. A
7. A
8. C
9. A
10. B

GLOSSARY

Adapter A device with a 3-slot receptacle for an appliance plug and a 2-prong plug for inserting into a 2-slot outlet.

AL A rating indicating that breakers are approved for use with aluminum branch circuit wiring.

Alternating current An electrical current, used in homes, in which electrons move back and forth at a frequency of 60 cycles per second.

Ampacity The number of amps that can safely pass through a given conductor.

Amperage See *Current*.

Amperage rating In home inspection, the size of amp service supplied to a home. For example, a 100-amp or 200-amp service.

Amps The unit of measure for current.

Anti-oxidant compound A grayish paste applied to aluminum wiring connections to prevent aluminum oxide from forming.

Bonding Electrically connecting two or more conductive items together and to the grounding system.

Branch circuit wiring That portion of the electrical system that runs from the electrical panel to outlets, switches, and fixtures in the home.

Breaker See *Circuit breaker*.

British Thermal Unit A unit of measure of heat output. Abbreviation *BTU*.

Busbar A conductor bar that provides connections and power for fuses and breakers and thus to circuits.

BX cable A metal armored cable used in branch circuit wiring which carried only a hot and neutral wire until the 1950's, but since also carries a ground wire.

Circuit A complete path of an electric current.

Circuit breaker An overload protection device that opens a circuit and stops the flow of electricity when the circuit overloads.

CO/ALR A rating indicating that breakers, connectors, switches, and outlets are approved for use with aluminum branch circuit wiring.

Conductor A material that offers a low resistance to an electric current flowing through it. Also, the wire used in the home's electrical system as in the grounding conductor, service conductor, and so on.

Conduit A metal cable, rigid or flexible, used in branch circuit wiring, which can carry two hot wires, a neutral, and a ground wire.

COPALUM Special crimp connectors that are used in connecting aluminum branch circuit wiring to copper at each outlet, switch, and junction.

Copper pigtailing A wire nut method used in connecting aluminum branch circuit wiring to copper at each outlet, switch, and junction.

Current The flow of electricity that results when the electromotive force is applied across a given resistance. Current, also called amperage, is measured in amps.

Dedicated circuit A circuit provided for the sole use of a large appliance such as the refrigerator or dishwasher.

Dielectric connector A type of connector used between two dissimilar metal pipes that prevents metal-to-metal contact between the pipes and stops the flow of electricity along the pipes.

Double tapping Adding wires at fuses or breakers for the purpose of adding more circuits to the electrical system. Not allowed, except with breakers approved for double tapping.

Drip loop: Slack in electrical wires at the masthead which prevent water from running into the conduit.

Electrical meter The meter provided by the utility company that measures home electrical use in watts.

Electromotive force The force that drives a current of electrons through a given resistance. Electromotive force, also called voltage, is measured in volts.

Fuse An overload protection device that melts and opens a circuit when the circuit overloads. May be cartridge or screw-in types. Type D fuses are time delay, type P is sensitive to heat buildup between the fuse and fuse holder, types S and C prevent the wrong size fuse fitting into the fuse holder.

GFCI An abbreviation for a ground fault circuit interrupter, a monitoring device that will trip after a ground fault is detected, stopping the flow of electricity in a circuit.

GFCI outlet An outlet which has a monitoring device that will trip the circuit when a ground fault is detected.

GFCI tester A testing device used to test the operation of GFCI outlets and 3-slot outlets for polarity and grounding.

Grounded outlet An outlet wired with a ground wire that is connected to the grounding system.

Grounding The process of electrically connecting electrically conductive items to the earth by which means excess electrical current is absorbed into the ground.

Grounding conductor The wire used to ground the home's electrical system.

Grounding system The means by which a home's electrical system is grounded, typically by connecting a ground wire to plumbing pipes or to a metal rod buried in the earth.

Hot wire An electrically charged wire, carrying a charge of 120 volts in home circuits.

Insulator A material that offers a high resistance to an electric current flowing through it.

Junction box A covered metal box used to protect connections or junctions in an electrical circuit.

Knob-and-tube wiring Old branch circuit wiring using ceramic knobs to secure wire to surfaces and tubes to pass wires through framing members.

Knockout The fuse holder. An open knockout is one without a fuse in it.

Load A light or appliance that uses electricity on a circuit.

Low voltage system A lighting system featuring boxes of relays, unusual switches, and one or more switching panels.

Main disconnect Fuse(s) or breaker(s) that alone or together stop the entire flow of electricity to the home. Located in a service box or the main panel.

Main overload protection device *See Main disconnect.*

Main panel A metal box holding overload protection devices and/or disconnects for the home's electrical circuits.

Multiple tapping See *Double tapping* and *Tapping before the main.*

National Electrical Code A national standard for electrical installations published by the National Fire Protection Association. Abbreviation *NEC.*

Neon bulb tester A testing device used to test 2-slot and 3-slot outlets for polarity and grounding.

Neutral busbar A conductor bar that provides connections for the neutral service conductor, neutral, circuit ground wires, and the grounding conductor.

Neutral wire A wire with no electrical charge (where equal and opposite currents cancel each other out) that provides a return path for electricity in a circuit.

Ohms The unit of measure for resistance.

Open ground Where an outlet is not grounded.

Outlet An electrical connecting device providing receptacles into which appliances can be plugged for power.

Overfusing Using a fuse on a circuit that is larger than the wire's capacity. Not allowed.

Overload protection device A fuse or breaker which will break the circuit when it overloads.

Polarized outlet An outlet with a large neutral slot and a smaller hot slot for plugging in appliances with large and small prongs.

Power The heat produced by the flow of current through a given resistance. Power is measured in watts, kilowatts, and BTU's.

Pull-out fuse block A handled block containing cartridge fuses that can be pulled out to stop the flow of electricity.

Resistance The opposition offered by a material when a current passes through it. Resistance is measured in units called ohms.

Reverse polarity Where the hot wire is wired to the large slot in an electrical outlet and the neutral wire is wired to the small slot, the opposite of how it should be done.

Rigid metal conduit See *Conduit*.

Romex cable A non-metallic sheathed cable used in branch circuit wiring which carries a hot, a neutral, and a ground wire.

Service box A metal box, separate from the main panel, that contains the main disconnect.

Service conductors The wires bringing electricity to the home.

Service drop Overhead wires bringing the electrical service to the home.

Service entrance The portion of a home's electrical system from the utility pole to the home's main disconnect.

Service lateral Underground wires bringing the electrical service to the home.

Single-bus panel An electrical panel with a single pair of busbars for either 120-volt or 240-volt circuits.

Split-bus panel An electrical panel with two or more pairs of busbars. The upper busbar is for 240-volt circuits, one of which provides power to the lower busbar. The lower busbar is for 120-volt circuits.

Split outlet An outlet wired with two hot wires and a neutral, which provide two separate circuits to the outlet, plus the ground wire for grounding purposes.

Subpanel A panel connected to a main panel for the purpose of providing more circuits and better distribution of electricity to the home.

Tapping before the main Adding wires before or at the main disconnect for the purpose of providing more circuits to the electrical service. Not allowed.

Voltage See *Electromotive force*.

Voltage rating In home inspection, the size of volt service supplied to the home. For example, a 120-volt or 240-volt service.

Volts The unit of measure for electromotive force.

Watts The unit of measure of power. One watt is equal to 3.4

INDEX

A Practical Guide to Inspecting Program
Study Unit Five, Inspecting Electrical

Student Name: _____ Date: _____

Address: _____

Phone: _____ Email: _____

After you have completed the exam, mail *this exam answer page* to American Home Inspectors Training Institute. You may also fax in your answer sheet. You will be notified of your exam results.

Fill in the box(s) for the correct answer for each of the following questions:

1. A☐ B☐ C☐ D☐	24. A☐ B☐ C☐ D☐	47. A☐ B☐ C☐
2. A☐ B☐ C☐ D☐	25. A☐ B☐ C☐ D☐	48. A☐ B☐ C☐ D☐
3. A☐ B☐ C☐ D☐	26. A☐ B☐ C☐ D☐	49. A☐ B☐ C☐ D☐
4. A☐ B☐ C☐ D☐	27. A☐ B☐ C☐ D☐	50. A☐ B☐ C☐ D☐
5. A☐ B☐ C☐ D☐	28. A☐ B☐ C☐ D☐	
6. A☐ B☐ C☐ D☐	29. A☐ B☐ C☐ D☐	
7. A☐ B☐ C☐ D☐	30. A☐ B☐ C☐ D☐	
8. A☐ B☐ C☐ D☐	31. A☐ B☐ C☐ D☐	
9. A☐ B☐	32. A☐ B☐ C☐ D☐	
10. A☐ B☐ C☐ D☐	33. A☐ B☐ C☐ D☐	
11. A☐ B☐ C☐ D☐	34. A☐ B☐ C☐ D☐	
12. A☐ B☐	35. A☐ B☐ C☐ D☐	
13. A☐ B☐ C☐ D☐	36. A☐ B☐ C☐ D☐	
14. A☐ B☐ C☐	37. A☐ B☐ C☐ D☐	
15. A☐ B☐ C☐ D☐	38. A☐ B☐ C☐ D☐	
16. A☐ B☐ C☐ D☐	39. A☐ B☐ C☐	
17. A☐ B☐ C☐ D☐	40. A☐ B☐ C☐ D☐	
18. A☐ B☐	41. A☐ B☐ C☐ D☐	
19. A☐ B☐ C☐ D☐	42. A☐ B☐ C☐ D☐	
20. A☐ B☐	43. A☐ B☐ C☐	
21. A☐ B☐ C☐	44. A☐ B☐ C☐	
22. A☐ B☐	45. A☐ B☐ C☐	
23. A☐ B☐ C☐	46. A☐ B☐	